Vanished Species

The Cost of Extinction

and What We Can Do to Stop It

Table of Contents:

Preface

It is an unsettling truth that we stand on the brink of a profound ecological crisis. Across the globe, habitats are shrinking, oceans are warming, forests are retreating, and countless species are disappearing—often before we even know they exist. This book was born out of a deep concern for this situation and the desire to illuminate the often-unseen connections between human activity and the extinction of species. Our aim is not merely to document the disappearance of wildlife, but to trace the tapestry of causes—hunting, habitat loss, pollution, invasive species, climate change—that bind all living beings in a complex and fragile web of life.

For many, the word "extinction" conjures images of dinosaurs vanishing in the prehistoric past. Yet extinction is neither confined to ancient history nor restricted to iconic megafauna. It continues in our own times, quietly and persistently, as modern civilizations expand, exploit resources, and alter natural landscapes at an unprecedented rate. The "Sixth Mass Extinction," a term used by many scientists, indicates that current extinction rates are drastically higher than normal background levels. While previous mass extinctions were triggered by volcanic eruptions, asteroids, or other natural cataclysms, this time the driving force is largely anthropogenic—caused by us. Through deforestation, overfishing, pollution, greenhouse gas emissions, and the introduction of invasive species, we humans have become a planetary force. The deeper one investigates these factors, the clearer it becomes that each extinct species tells a story of imbalance, where nature's resilience has been stretched to breaking point.

I can vividly recall the moment that sparked my personal dedication to this topic. Several years ago, I visited an exhibit in a small natural history museum. Tucked in a modest corner was a display about the Quagga—an extinct subspecies of zebra once found in South Africa. I was captivated by its distinct pattern: bold stripes in the front merging into a brownish coat in the rear. Reading about its extinction in the late 19th century, largely due to hunting by colonial settlers, felt like an unexpected personal loss. It wasn't just the Quagga's story that left me unsettled, but also the pattern of human behavior it represented. From that day on, each account of an extinct animal became not just a historical fact but a call to reflect on how our actions impact the planet we share.

Over time, I discovered that the Quagga's story is echoed by many others. The charismatic Dodo, the imposing Barbary Lion, the gentle Tecopa Pupfish—all victims of pressures linked to human activity. Yet, these are but a fraction of a far greater number of species lost forever. Like the Quagga, many of these animals fell victim to overhunting or habitat encroachment, while others succumbed to introduced diseases or invasive predators. In each case, beneath the scientific descriptions and the dates of discovery and extinction, lies a deeper narrative of how our societal choices—from where we settle to what we consume—reverberate through ecosystems worldwide.

The structure of this book reflects the variety of life forms that have vanished: mammals, birds, reptiles, amphibians, fish, and insects. This categorization is not meant to pigeonhole nature—indeed, life defies neat classifications—but rather to illustrate the breadth of biodiversity impacted by the forces we have unleashed. We begin by examining extinct mammals (Chapter 1), such as the Tarpan, Javan Tiger, and Pyrenean Ibex. This leads us into the stories of extinct birds (Chapter 2), including the iconic Dodo and the Passenger Pigeon, once so abundant that they darkened the skies of North America in massive flocks. We then delve into extinct reptiles (Chapter 3)—a sobering look at the Giant Tortoises of the Seychelles and the famed Pinta Island Tortoise from the Galápagos. Next, we move to extinct amphibians (Chapter 4), shining a light on creatures like Darwin's Frog and others lost to climate change and virulent fungal diseases. Our journey continues with extinct fish (Chapter 5), many of which disappeared as rivers were dammed, lakes polluted, and coastal waters overfished. Finally, we turn to extinct insects (Chapter 6), often overlooked but vital pollinators and recyclers in their respective habitats. In each chapter, we explore the primary causes—be they hunting, habitat fragmentation, pollution, or invasive species—that have driven these extinctions.

Following the species-specific chapters, the book shifts toward a wider ecological perspective. In Chapter 7, we discuss the impact of these losses on ecosystems: the domino effects that ripple outward when one species vanishes, altering entire food webs and nutrient cycles. This is where the gravity of mass extinctions becomes most apparent. The disappearance of a single pollinator species can threaten entire plant communities, while the loss of a "keystone predator" can spark population explosions in certain prey, ultimately harming vegetation, water quality, and soil composition. Such cascades remind us that life on Earth is interwoven beyond our immediate perception. We are part of a grand puzzle, where each piece plays a role in the stability of the whole.

Yet, for all the dire implications, this book is not intended to be a eulogy. Rather, it is a wake-up call. In Chapter 8, we examine strategies and actions that can help halt—if not reverse—current trends. Conservation and protection of endangered species, habitat restoration, and the enforcement of international treaties like CITES (Convention on International Trade in Endangered Species) are crucial. Education, too, plays a pivotal role. By raising awareness in schools, universities, and community groups, we foster a sense of stewardship that transcends generations. The final chapter also invites individual reflection: how might each of us, in our daily routines, contribute to sustainability? Whether through reducing waste, supporting ethical farming practices, or volunteering with local conservation groups, there is always a meaningful step we can take.

To better understand how we reached this predicament, it is essential to place these extinctions within a broader historical framework. We did not arrive at this moment overnight. Humanity's relationship with nature has evolved over millennia, from the early days of small hunter-gatherer communities to the modern era of industrial capitalism and global trade. Early human societies may have hunted some species to local or global extinction—such as woolly mammoths and giant ground sloths—often without realizing the long-term consequences. With the advent of agriculture, the ability to radically transform landscapes for crops and livestock expanded our ecological footprint. The Industrial Revolution accelerated this shift exponentially, with fossil fuel consumption enabling mass production, rapid transportation, and an ever-growing appetite for natural resources. Today's global economy further amplifies these effects, connecting regions worldwide in complex supply chains that mask the immediate environmental damage of our consumption.

In writing this book, we were mindful that statistics and grim details alone can lead to a sense of helplessness. Indeed, confronting the scale of species loss can be overwhelming. Thus, while we have not shied away from illustrating the stark realities—populations reduced to a handful of survivors, animals forced into smaller and smaller fragments of habitat—we have equally emphasized narratives of hope and resilience. Nature has an incredible capacity to rebound when given the chance. Case studies of certain species once deemed functionally extinct, only to be rediscovered or reintroduced through dedicated conservation efforts, demonstrate that extinction is not always a foregone conclusion. Moreover, the success of reforestation projects, wetland restoration, and captive breeding programs shows us that with sufficient political will, community involvement, and scientific expertise, we can make a difference.

Another recurring theme in these pages is the moral and ethical dimension of extinction. Some might argue that extinction is part of Earth's evolutionary process, a natural occurrence that has always happened. While it is true that species have emerged and disappeared over billions of years, the current rate and magnitude of loss is anything but natural. Far from reflecting the slow pace of natural selection, these modern extinctions are often abrupt and concentrated in biodiverse hotspots. They coincide with periods of intense human activity—whether logging, mining, pollution, or large-scale agriculture. A troubling question emerges: what does it say about our species that we are erasing so many forms of life? Are we fulfilling our role as stewards of this planet, or neglecting it?

In grappling with these questions, we recognize that the responsibility for extinctions is unequally distributed. Communities living in harmony with their local ecosystems are often overshadowed by large multinational corporations, regulatory gaps, or simply the inertia of global markets. Poverty, too, can drive environmental degradation—people who lack basic needs understandably prioritize immediate survival over long-term conservation. This context underscores the complexity of devising solutions. Effective conservation efforts require not only ecological expertise but also cultural sensitivity, social justice, and inclusive economic development. Collaborations between governments, NGOs, indigenous communities, local businesses, and international agencies become indispensable.

Throughout the writing process, we have also reflected on the extraordinary diversity of life that still exists. From tropical rainforests to polar ice caps, from the open ocean to volcanic islands, Earth teems with life forms that continually astound scientists and naturalists alike. Even as we mourn the loss of the Northern White Rhinoceros or the Xerces Blue Butterfly, we should remember that countless wonders remain. The same fundamental processes that gave rise to these now-lost species continue to shape the biosphere. Evolution, adaptation, and resilience are ongoing. Our task is to ensure that we do not push these processes beyond nature's capacity to renew itself.

In the pages to come, you will encounter stories of awe and melancholy, of scientific discovery and tragic finality. You will read about once-thriving flocks of Passenger Pigeons reduced to zero, about the tragic last known Thylacine in a Tasmanian zoo, and about the systematic eradication of the Caspian Tiger in vast stretches of Eurasia. You will discover how invasive species, like rats and cats introduced to fragile island environments, wreaked havoc on birds that had evolved for millennia without ground predators. You will come across the heartbreak of amphibians decimated by chytrid fungal infections, largely spread through global trade. And you will learn how the wholesale conversion of wetlands to farmland spelled doom for fish like the Tecopa Pupfish, an evolutionary marvel adapted to the extreme conditions of the American desert.

But you will also see the unwavering determination of conservationists, local communities, and scientists who strive to protect what remains. By preserving habitats, establishing reserves, creating wildlife corridors, and regulating hunting, they have kept certain species from following the same path to oblivion. They have proven that knowledge, passion, and concerted effort can counterbalance even centuries of destructive practices. In amplifying these efforts, we hope to inspire readers to see themselves as part of the solution. After all, every individual choice—from how we vote, to what we buy, to the conversations we have with friends and family—shapes a collective outcome.

This book, then, is an invitation. An invitation to bear witness to the stories of species lost and threatened, and to reflect on their broader significance for the planet's future. An invitation to recognize that we stand at a pivotal moment in Earth's history—one in which our actions, whether deliberate or negligent, will echo through the ages. If you feel compelled to grieve these vanishings, that is natural. Let that grief be the seed of a deeper understanding of how intricately and delicately life is woven together, and how extraordinary our responsibility is to preserve it.

Ultimately, each of us has a role to play, whether as educators, parents, scientists, artists, policymakers, or simply concerned citizens. We may not solve every challenge, but we can champion initiatives that restore habitats, protect remaining wildlife, and balance human needs with ecological integrity. Our hope is that by reading these chapters—by understanding the tragic destinies of the Uro, the Barbary Lion, the Pinta Island Tortoise, the Elephant Bird, and so many others—you will gain not just facts but also empathy and motivation.

Yes, we have lost much. But we are also at the cusp of a new awareness, where global collaboration, technological innovation, and a growing moral consciousness can converge to forge a different path. Instead of being defined by the extinctions we cause, we can choose to be remembered as the generation that halted the decline, that saw beyond short-term profits and recognized the intrinsic value of every living thing. We can cultivate the humility and the wisdom that come from acknowledging our dependence on Earth's biodiversity. And in doing so, we can pass on a richer, more vibrant planet to those who follow.

The pages ahead present a meticulous yet heartfelt account of what has been lost—and what we still stand to lose—if we continue on our current trajectory. May this journey through time and across continents expand your perspective, inspire dialogue, and above all, mobilize action. Our shared future depends on it. Let us begin.

CHAPTER 1: EXTINCT MAMMALS

Mammals have long captured human imagination and reverence, from the awe-inspiring lions and tigers that roamed vast continents to the humble wild cattle that once roamed Europe's forests. Their size, complexity, and, in many cases, behavioral similarities to humans have made mammals emblematic figures in the story of extinction. In this chapter, we will examine notable cases of extinct mammals, focusing on their evolutionary history, ecological roles, cultural significance, and the primary drivers that led to their disappearance. By weaving together these narratives, we hope to illustrate both the scale of human impact and the multifaceted nature of biodiversity loss.

1.1 Uro (Bos taurus primigenius)

Introduction and Description

The Uro, also known by its scientific name Bos taurus primigenius, was a wild ancestor of modern domestic cattle. Renowned for its imposing stature and impressive horns, the Uro once inhabited large swaths of Europe, Asia, and North Africa. Historical accounts describe it as having a massive, muscular frame, with bulls often reaching up to 1.8 meters at the shoulder and weighing over 1,000 kilograms. Their coat color typically ranged from dark brown to black in males, while females were somewhat lighter in tone. The horns, curving forward and then inward, were formidable weapons against predators and served as visual displays during intraspecific conflicts.

Evolutionary Significance

Genetically, the Uro plays a pivotal role in the lineage of domesticated cattle (Bos taurus). Archaeological evidence suggests that domestication events occurred independently in different regions—most notably in the Fertile Crescent and the Indian subcontinent—giving rise to multiple cattle lineages that still show genetic traces of the Uro. By studying ancient Uro DNA, researchers have gleaned insights into how domestication changed not only the animal's phenotype (size, horn shape, temperament) but also its physiology and reproductive patterns.

Historical Range and Habitat

During the Middle Ages, Uro populations could be found in dense deciduous forests, wetlands, and open woodlands across Europe. Their preferred habitats offered ample grass, shrubs, and water sources for grazing and wallowing. As large herbivores, Urus helped shape the vegetation structure by grazing, dispersing seeds through their dung, and creating pathways in dense undergrowth. This "ecological engineering" had a ripple effect on biodiversity, benefiting smaller mammals, birds, and insects.

Cultural and Mythological Importance

The Uro held a special place in the cultural and mythological tapestry of many early civilizations. Cave paintings across Europe—such as those in Lascaux, France—depict them in vivid detail, highlighting their significance to prehistoric hunter-gatherers. In Roman times, they were sometimes captured for use in gladiatorial games, as their size and temper made them formidable opponents. Over centuries, the Uro became intertwined with stories of strength and wildness, symbolizing an untamed force of nature in folklore from Poland to the Caucasus region.

Causes of Decline

Overhunting: As human populations expanded, the Uro was hunted relentlessly for its meat, hide, and horns. In some areas, they were also targeted for sport by nobility, who considered the hunt a prestigious pastime.

Habitat Loss: Agricultural expansion and widespread deforestation fragmented the Uro's habitat. Where once large tracts of forest offered ample space, farmland and settlements squeezed the remaining populations into smaller, isolated pockets.

Domestication Pressures: While domestication itself did not drive the Uro to extinction, the shift from wild herds to domesticated cattle reduced the genetic diversity of the original wild populations. Over time, fewer truly wild Uros remained outside managed lands.

Final Demise and Last Records

By the 16th century, Uro populations were restricted to remote parts of Eastern Europe, most notably in the forests of Poland. Royal decrees attempted to protect them, but these efforts were largely symbolic and insufficient. The last recorded Uro is believed to have died in the Jaktorów Forest in Poland in 1627—likely from disease, though the broader context includes a centuries-long decline caused by overhunting and habitat encroachment.

Legacy and "Breeding Back" Efforts

In the 20th century, some European conservationists attempted to "recreate" the Uro by selectively breeding domestic cattle that retained certain Uro-like traits, such as large horns and dark coats (e.g., the Heck cattle project). While these modern breeds offer a semblance of the animal's appearance, they are genetically distinct from true Uros. Nevertheless, these efforts underscore a growing recognition of the Uro's ecological and cultural importance, and a desire to restore wild cattle to European landscapes—even if only as ecological proxies rather than genuine resurrected Uros.

1.2 Pyrenean Ibex (Capra pyrenaica pyrenaica)

Introduction and Physical Characteristics
The Pyrenean Ibex was a subspecies of the Spanish Ibex (Capra pyrenaica), uniquely adapted to the rugged mountain environments of the Pyrenees, the natural

border between Spain and France. Adult males boasted large, backward-curving horns, which could measure up to 75 centimeters in length and featured distinctive ridges. Their coats were thick and coarse, especially in winter, aiding survival in frigid alpine conditions. Females, smaller and less dramatically horned, traveled in herds with their offspring, while males often led more solitary lives, joining herds during the mating season.

Habitat and Ecological Role

Occupying rocky slopes and high-altitude meadows, Pyrenean Ibex fed primarily on grasses, lichens, and hardy shrubs. By grazing on alpine vegetation, they prevented overgrowth and helped maintain open patches, which benefited various flora and fauna. In turn, they fell prey to natural predators like wolves and eagles, although human hunting rapidly became a more significant threat.

Population Decline and Causes

Intensive Hunting: Historically prized as a trophy for their horns and meat, Pyrenean Ibex were hunted to near oblivion by the mid-20th century. In some regions, hunting was unregulated, and even after protective measures were introduced, poaching continued.

Habitat Fragmentation: Infrastructure development—roads, ski resorts, and dams—fragmented their alpine habitats. With fewer safe corridors for movement, populations became isolated, diminishing their genetic diversity and resilience.

Competition with Domestic Livestock: As humans introduced sheep and goats to high-altitude pastures, competition for limited resources intensified. Additionally, livestock could carry diseases that spread to wild ibex herds.

Conservation Efforts and Cloning Attempt

By the early 1990s, the Pyrenean Ibex was critically endangered, with only a small number of individuals left. Conservation programs aimed to protect the last herds, but these efforts came too late. The final known female died in 2000, officially marking the subspecies' extinction. However, in 2003, a groundbreaking cloning experiment briefly "resurrected" the Pyrenean Ibex using DNA from a preserved skin sample. A cloned female survived just minutes after birth due to respiratory failure, making it the first taxon in history to become "un-extinct" and then extinct once more. This event sparked global debates about de-extinction technologies, ethics, and the feasibility of sustaining a cloned population without a robust genetic pool or suitable habitat conditions.

Symbolic Lessons

The Pyrenean Ibex case highlights the delicate balance of mountain ecosystems. It illustrates how even adapted, resilient species can succumb to a combination of overexploitation and habitat pressures. The cloning experiment, meanwhile, forces us to reconsider our relationship with nature: should we rely on scientific breakthroughs to correct past mistakes, or should we prioritize in-situ conservation before species reach the brink?

Genetic Legacy in Related Species

Although Capra pyrenaica pyrenaica is gone, closely related subspecies like the Southeastern Spanish Ibex (C. p. hispanica) and the Gredos Ibex (C. p. victoriae) still survive. Conservationists analyze their genetics to gain a clearer picture of the Pyrenean Ibex's lost genetic traits. Insights from these efforts could guide future breeding programs to prevent similar fates for the remaining ibex subspecies.

1.3 Javan Tiger (Panthera tigris sondaica)

Origins and Distinguishing Traits
The Javan Tiger was one of the three tiger subspecies native to the Indonesian archipelago (the other two being the Sumatran Tiger and the extinct Balinese Tiger). Often described as smaller and darker than mainland tigers, the Javan Tiger featured closely spaced stripes and a distinctly narrow face. Its habitat once spanned the lush rainforests and grasslands of Java, an island known for its volcanic soils, dense human populations, and rich biodiversity.

Habitat and Ecological Role

Java's tropical climate supported a mosaic of forests, marshlands, and mountain slopes. The Javan Tiger played a critical role as an apex predator, keeping herbivore populations in check. Deer, wild boar, and primates formed a large part of its diet, and by regulating these species' numbers, tigers helped maintain balanced vegetation dynamics. In healthy ecosystems, large predators are keystone species, exerting profound influence over trophic cascades and overall biodiversity.

Human Encroachment and Agricultural Pressures
Over centuries, Java's human population grew exponentially. By the early 20th century, the island faced intense deforestation as vast tracts of rainforest were converted to rice paddies and plantations (teak, rubber, coffee, and later palm oil). This rapid agricultural expansion restricted the Javan Tiger to fragmented forest pockets, where contact with humans became more frequent. Livestock predation sometimes spurred

retaliatory killings, while trophy hunting and the trade in tiger parts, fueled by demand in traditional medicine, placed additional pressure on dwindling populations.

Conservation Measures and Decline

In the 1940s and 1950s, Javan authorities recognized the tiger's precarious status, establishing several protected areas like Ujung Kulon National Park. However, these refuges proved insufficient and too fragmented. The Indonesian government's focus on economic development and population resettlement projects sometimes conflicted with conservation goals. As the habitat shrank, tiger populations became genetically isolated, increasing vulnerability to diseases and inbreeding.

The last reliable sightings date to the late 1970s or early 1980s in eastern Java, specifically in the Meru Betiri region. Some anecdotal reports persisted into the 1990s, but these remain unverified. Most experts concur that the Javan Tiger is almost certainly extinct.

Cultural Significance and Myths

Tigers in Javan and wider Indonesian folklore often represent strength, courage, and supernatural power. The disappearance of the Javan Tiger signaled not just a loss of an apex predator but a rupture in cultural narratives. Local traditions sometimes considered the tiger a guardian of the forest, a notion that resonates with ecological science emphasizing the predator's role in maintaining healthy ecosystems.

Lessons Learned

Population Pressure: Java is among the world's most densely populated islands, highlighting the challenge of reconciling human needs with wildlife conservation.

Habitat Fragmentation: Even when tigers received legal protection, the continued clearance of forests undermined such measures.

Importance of Community Engagement: Conservation strategies that fail to involve local communities often fall short, as successful initiatives require addressing human-wildlife conflicts and offering livelihoods compatible with protection efforts.

Potential for Rewilding?

Some conservationists have speculated about reintroducing tigers to Java if suitable habitats can be rehabilitated. However, given ongoing deforestation and the precarious state of other Indonesian tiger populations (like the Sumatran Tiger), such a project remains highly unlikely. The Javan Tiger's story stands as a stark reminder of how quickly a large predator can vanish when faced with relentless anthropogenic pressures.

1.4 Caspian Tiger (Panthera tigris virgata)

Range and Physical Characteristics
The Caspian Tiger once roamed an expansive territory stretching from modern-day Turkey and the Caucasus to Central Asia and parts of northwestern China. Often conflated with the Siberian Tiger, the Caspian Tiger was slightly smaller but shared a similarly robust build and thick fur adapted to colder climates. Known to inhabit riparian forests and reed beds along rivers like the Amu Darya and Syr Darya, this subspecies thrived in habitats that offered ample cover and game, such as wild boar and deer.

Historical Accounts

Ancient texts and archaeological evidence suggest the Caspian Tiger had cultural significance in Persia and surrounding regions. Royals hunted them for sport, and their pelts were highly valued. In some folklore, the tiger symbolized both regal authority and untamed wilderness.

Key Drivers of Extinction

Land Use and Agriculture: During the 19th and early 20th centuries, the Russian Empire, and later the Soviet Union, initiated massive land reclamation and irrigation projects. Wetlands and reeds were cleared to make way for cotton plantations, drastically shrinking the tiger's habitat.

Hunting Campaigns: The Caspian Tiger was treated as a pest in many Soviet campaigns aimed at "taming" frontier regions. Bounties were offered, leading to organized hunts that decimated populations.

Decline of Prey: As wild boar and deer populations declined due to overhunting and habitat alteration, tigers found it increasingly difficult to secure sufficient prey. Nutritional stress further reduced their reproductive success.

By the mid-20th century, the Caspian Tiger's range had become patchy. The last verified sightings date to the 1970s in remote areas of the Tian Shan and Pamir Mountains. Although there were occasional, unconfirmed reports afterward, most scientists now consider the Caspian Tiger extinct.

Genetic Connections with the Amur Tiger Intriguingly, genetic research reveals that the Caspian Tiger shared a close relationship with the Amur (Siberian) Tiger (Panthera tigris altaica). Some scientists propose that the Caspian Tiger's ancestors colonized Central Asia from eastern regions, meaning the two subspecies diverged relatively recently. This genetic link has prompted discussions about whether reintroduction using Amur Tigers could restore a "Caspian-like" tiger to parts of its former range, particularly the riverine forests of Central Asia. However, such rewilding schemes face the same challenges as the Javan Tiger scenario: habitat loss, regional conflicts, and competition for land and resources.

Conservation Lessons

The Caspian Tiger's story underscores the dangers of large-scale agricultural projects that ignore environmental impacts. It also highlights the intersection of conservation and geopolitics—decisions made by empires and states to transform landscapes can have irreversible consequences on wildlife. Today, efforts to protect the Amur Tiger in Russia's Far East serve in part as a homage to the lost Caspian Tiger, exemplifying how crucial it is to preserve viable habitats before species are pushed beyond the point of recovery.

1.5 Barbary Lion (Panthera leo leo)

Historical Significance

The Barbary Lion, also known as the Atlas Lion, once roamed across North Africa, from Morocco's Atlas Mountains to Egypt. Celebrated for its majestic mane—often extending beyond the shoulders—this lion was famously used in Roman arenas, symbolizing both imperial power and Africa's wild riches.

Appearance and Ecology

Barbary Lions were believed to be among the largest lion subspecies, with some males reportedly weighing over 200 kilograms. The mane could be notably thick and dark, an adaptation to the cooler highland climates. While their precise prey spectrum varied by region, Barbary Lions primarily hunted large ungulates such as

Barbary stags and gazelles. Their territory included mountainous woodlands and semi-desert areas, reflecting North Africa's diverse topography.

Human Pressures and Decline

Overhunting and Capture: Roman emperors captured Barbary Lions for gladiatorial combat. Centuries later, colonial hunters and local rulers alike sought them as trophies. The repeated exploitation caused lion numbers to plummet.

Habitat Modification: Deforestation for timber and farmland reduced prey availability, forcing the lions into smaller pockets of suitable habitat.

Conflict with Pastoralists: Lions that preyed on livestock were systematically killed. As human populations expanded in North Africa, encounters with lions became more frequent, often ending in the lion's demise.

By the mid-20th century, wild Barbary Lions were extremely rare. Sightings in the Atlas Mountains during the 1920s and 1930s mark some of the last credible records. The subspecies is presumed extinct in the wild by the early 1960s, though a few individuals may have survived in remote mountain regions slightly beyond that date.

Legends of Survival and Captivity

There have been persistent rumors of remnant Barbary Lions in Moroccan or Algerian enclaves, but these remain largely unsubstantiated. Some captive lions in zoos claim lineage from Barbary Lions once held by royal courts in Morocco, yet genetic analyses show that most are heavily mixed with other African lion subspecies. Certain conservationists have advocated selectively breeding captive lions with partial Barbary ancestry to restore this iconic lineage, but the feasibility and ethical dimensions remain debated.

Cultural Echoes

The Barbary Lion looms large in North African folklore and heraldry—Morocco's national football team is nicknamed "The Atlas Lions." Across centuries, these lions served as emblems of power and wilderness. Their disappearance thus represents more than just an ecological loss; it symbolizes the fading connection between modern societies and the wild predators they once revered.

Learning from the Barbary Lion

The Barbary Lion exemplifies how a charismatic top predator can disappear when hunted relentlessly and deprived of its habitat. The subspecies' fate also foreshadows the risks facing other lion populations today, many of which confront similarly dire pressures across the African continent. Conservation strategies for existing lion populations—such as mitigating human-wildlife conflict and establishing protected corridors—might prevent other regions from witnessing a similar tragedy.

1.6 Northern White Rhinoceros (Ceratotherium simum cottoni)

Taxonomy and Distribution

The Northern White Rhinoceros is (or was) one of two subspecies of the White Rhinoceros, the other being the Southern White Rhinoceros (C. s. simum). Historically, the Northern subspecies inhabited parts of Uganda, Sudan, the Central African Republic, and the Democratic Republic of Congo, roaming savannas and grasslands with abundant water sources.

Distinguishing Features

White rhinoceroses are characterized by their wide, square lips adapted for grazing. The "white" in their name is likely a misinterpretation of the Dutch word "wijd," meaning "wide." Northern White Rhinos were marginally smaller and had subtle differences in skull shape compared to their southern counterparts. Nonetheless, both subspecies weigh between 1,600 and 3,600 kilograms, ranking them among the largest land mammals after elephants.

Catastrophic Decline

Poaching for Horn: Rhino horns have long been valued in some traditional medicine markets and as status symbols. By the late 20th century, rampant poaching fueled by high black-market prices decimated Northern White Rhino populations, reducing them from thousands to mere dozens.

Political Instability: Many Northern White Rhino habitats lay in conflict zones marked by civil wars and insufficiently funded conservation agencies. The breakdown of law and order in these areas facilitated poaching and hampered anti-poaching measures.

Loss of Genetic Diversity: As the population dwindled, inbreeding depression became a serious concern, further reducing the viability of remaining individuals.

The Last Survivors and Assisted Reproduction
By the early 2000s, only a handful of Northern White Rhinos survived in Garamba National Park (DR Congo). Conservationists relocated some to more secure locations, such as Ol Pejeta Conservancy in Kenya. Despite round-the-clock armed protection, these final individuals failed to breed naturally. As of 2021, only two females—Najin and Fatu—remained alive, both

incapable of carrying a pregnancy to term. An ambitious assisted reproduction program is exploring in-vitro fertilization using sperm from deceased males and perhaps surrogate Southern White Rhinos as gestational carriers.

Symbol of Modern Conservation Challenges

The plight of the Northern White Rhino is emblematic of conservation in conflict zones, illustrating how a species can approach the brink of extinction even when its biology and ecology are relatively well understood. The immense costs involved in safeguarding a handful of individuals—plus the advanced reproductive technologies required—prompt a deeper reflection: how can we prevent such crises before they unfold?

Reflections for the Future

Whether the Northern White Rhino can be "saved" through science remains uncertain. Even if the subspecies is ultimately lost, its story underscores the urgency of robust international cooperation. Nations, NGOs, and local communities must collaborate to ensure anti-poaching measures, sustainable tourism, and habitat protection are viable. Without systemic change, future species may follow in the Northern White Rhino's footsteps, necessitating even more desperate and expensive interventions.

1.7 Tarpan (Equus ferus ferus)

Wild Ancestors of Modern Horses
The Tarpan was a wild horse that roamed the steppes and woodlands of Eastern Europe and parts of Western Asia. It represents an essential link in the equine lineage, embodying traits that modern domesticated horses (Equus caballus) partially inherited. Tarpans stood around 130 to 140 centimeters at the shoulder, with a sturdy build, a dun-colored coat, and often a dorsal stripe down the back.

Habitat and Behavior

Preferring open grasslands and light forest edges, Tarpans were adapted to harsh, continental climates. They likely formed small herds led by dominant mares, with stallions vying for breeding rights. Their diet consisted of grasses, herbs, and, in winter, bark or shrubs when other forage was scarce. Known for agility and resilience, Tarpans could outrun predators like wolves, but human hunters posed an altogether different threat.

Path to Extinction

Overhunting and Captures: European peasants and nobles hunted Tarpans for meat, hides, and to eliminate competition for pasture with domestic livestock. Some Tarpans were captured and crossbred, diluting the wild gene pool.

Habitat Loss: As agriculture expanded, natural steppe habitats were plowed under for crops or converted into grazing lands for cattle and sheep.

Domestication: While the Tarpan itself was never broadly domesticated, the existence of domesticated horses reduced the perceived need to protect wild herds.

By the early 19th century, Tarpans survived only in fragmented pockets of Eastern Europe. The last documented wild Tarpan is believed to have died in the late 19th century, possibly around 1887 in Russia. Some captive individuals persisted in zoos until the early 20th century, but they were often hybrids, blurring the line between true Tarpans and domesticated horses.

Re-creation Projects

Similar to the Uro, various breeding programs—such as those involving the Konik horse in Poland—attempted to recreate a Tarpan-like phenotype. While these "Tarpan proxies" resemble the extinct wild horse in coat color and size, they are genetically distinct from true Tarpans. Nevertheless, these efforts have played a role in rewilding projects aimed at restoring natural grazing dynamics in parts of Europe.

Ecological Importance

Wild horses can shape grassland ecosystems by grazing, trampling vegetation, and dispersing seeds. Their movements prevent the landscape from turning into dense shrubland, maintaining open habitats favorable to a range of species. The loss of Tarpans thus removed a key herbivore from the European steppe ecosystem. Modern reintroductions of horse-like proxies seek to reestablish these ecological functions.

Cultural Perspectives

Throughout European history, the figure of the wild horse has symbolized freedom and natural power. In Slavic and Germanic mythologies, wild horses were imbued with spiritual significance, sometimes associated with fertility and warfare. The extinction of the Tarpan represents a break from that ancestral connection and a reminder that cultural admiration does not always translate into conservation.

1.8 Tasmanian Tiger (Thylacinus cynocephalus)

Unusual Marsupial Predator

Despite its name, the Tasmanian Tiger—also called the Thylacine—was neither a tiger nor a canine, though it superficially resembled both. A carnivorous marsupial, it had stripes across its back and a stiff tail reminiscent of a kangaroo. Indigenous to Australia, Tasmania, and New Guinea in prehistoric times, it retreated primarily to Tasmania following the arrival of dingoes in mainland Australia.

Ecology and Behavior

Tasmanian Tigers were nocturnal predators that fed on small mammals and birds. Their jaws could open remarkably wide, a unique adaptation among carnivorous marsupials. Living in dens or hollow logs, females carried up to four joeys in a pouch. Although solitary hunters, they occasionally worked in pairs or family units, especially when larger prey was targeted.

Human Conflict and Bounty Hunting

From the 1800s onward, European settlers in Tasmania blamed the Thylacine for livestock losses, particularly sheep. The government responded with bounties, incentivizing widespread killing. While some Thylacines likely did prey on vulnerable lambs, the scale of persecution far exceeded any realistic threat they posed. Habitat loss—through deforestation for agriculture and logging—further pressured their dwindling populations. By the early 20th century, numbers were critically low, confined mostly to remote highland forests.

Last Known Individual and Extinction

In 1933, a Thylacine was captured and sent to the Hobart Zoo, where it lived for three years. Named "Benjamin" (though its sex remains disputed), it died on September 7, 1936—just two months after the Tasmanian government had belatedly granted the species official protection. This tragic timing underscored the systemic failure to enact effective conservation measures sooner. Despite sporadic, unconfirmed sightings since then, no concrete evidence of surviving wild Thylacines has emerged.

Cultural Legacy

Indigenous Tasmanian communities revered the Thylacine as part of their ancestral landscapes. Post-colonial attitudes were often hostile, labeling it a pest or threat. In recent decades, however, the Tasmanian Tiger has gained iconic status in Australian conservation circles, symbolizing the fragility of island ecosystems and the devastating impacts of misguided persecution.

Scientific Interest and De-extinction

The Tasmanian Tiger's extinction has fueled ongoing fascination, prompting efforts to retrieve DNA from preserved specimens. While some researchers have contemplated "de-extinction" techniques—such as CRISPR gene editing—no breakthrough is imminent. Even if a viable genetic clone were produced, re-establishing a stable population in the wild would require intact habitats and a thorough understanding of the species' social structure and disease susceptibility.

Ethical Questions

Reviving the Tasmanian Tiger raises ethical dilemmas about resource allocation. Should significant funds be used to resurrect one species when hundreds of extant threatened species lack adequate protection? The Thylacine's story thus highlights both the allure and the complexities of advanced genetic interventions in conservation.

1.9 Quagga (Equus quagga quagga)

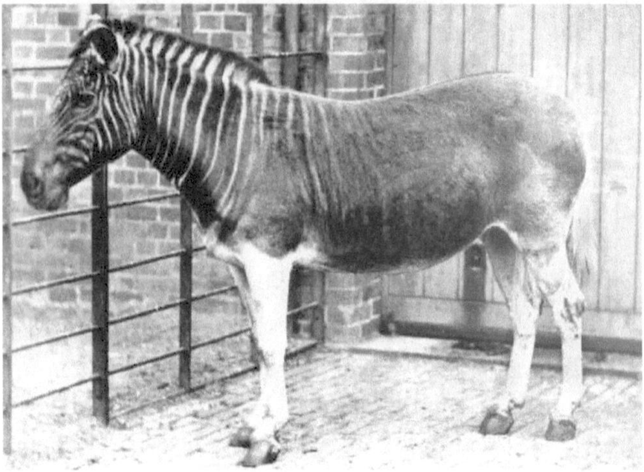

A Distinctive Subspecies of Zebra
The Quagga, native to South Africa's Cape region, presented a striking coat pattern. The front half bore bold zebra stripes, fading into a plain brownish coat toward the hindquarters. This unique appearance sparked curiosity among early European settlers, but also made the Quagga an easy target for trophy hunters.

Habitat and Range

Quaggas inhabited grasslands and arid savannas, forming herds that roamed alongside other plains game like blesbok and springbok. Grazing in mixed groups helped reduce predation risks from lions and other large carnivores. Yet, as European colonization expanded, competition with domestic livestock intensified, and

Quaggas were systematically hunted for their meat and hides.

Rapid Decline

During the 19th century, large-scale hunting campaigns decimated Quagga populations. Within a few decades, what had once been numerous herds all but vanished. The last known wild Quagga likely perished in the late 1870s, and the final captive individual died at the Amsterdam Zoo in 1883.

Scientific Rediscovery

At first, some naturalists considered the Quagga a separate species. Subsequent DNA analyses revealed that it was, in fact, a subspecies of the Plains Zebra (Equus quagga). This discovery showed that the zebra family tree is more fluid than once believed, with variations in striping patterns and coat color emerging relatively quickly in evolutionary terms.

Quagga Project

In the 1980s, a South African initiative called the Quagga Project commenced a selective breeding program using Plains Zebras with reduced striping. Over successive generations, some foals exhibited fewer stripes on their hindquarters, resembling historic Quagga pelts. While these "Rau Quaggas" bear a superficial resemblance, they remain genetically distinct from the original Quagga. Nonetheless, the project underscores a broader interest in "breeding back" certain phenotypes to restore lost ecological dynamics.

Ecological Role

As grazers, Quaggas were integral to the grassland ecosystem, influencing vegetation composition and providing prey for apex predators. Their disappearance, along with that of other large herbivores in certain regions, altered plant community structures and may have had cascading effects on other fauna.

Conservation Message

The Quagga's story is emblematic of the speed at which human actions—particularly overhunting—can eradicate a species. In less than a century, an animal once common enough to be largely overlooked vanished completely. Today, greater awareness of the interconnectedness of species and ecosystems could prevent such tragedies from recurring. Initiatives like the Quagga Project highlight both our fascination with restoring what has been lost and the moral imperative to protect remaining zebra subspecies from a similar fate.

1.10 Main Causes: Hunting, Poaching, Habitat Loss

Having explored these nine emblematic extinct mammals—from iconic predators like the Javan and

Caspian Tigers to the more obscure but ecologically pivotal Uro—several recurring themes emerge. Each extinction narrative reflects a unique blend of historical circumstances, yet they converge around three primary drivers: hunting, poaching, and habitat loss. These forces are often intertwined, amplifying one another until populations can no longer sustain themselves.

Hunting and Poaching

Subsistence vs. Commercial Hunting: Early human societies hunted for food and materials, sometimes pushing species to local extinctions. With the advent of global trade and advanced weaponry, however, hunting intensified. Animals like the Quagga were decimated by commercial hunts for hides, while tigers fell victim to a lucrative black market in skins and body parts.

Trophy Hunting: Large mammals, especially predators and megafauna, became prestigious trophies. Royal families, wealthy adventurers, and colonial administrators often engaged in hunts that devastated populations.

Bounty Systems: Government-endorsed bounties for perceived "pests" (such as the Tasmanian Tiger) spurred indiscriminate killing.

Habitat Loss and Fragmentation

Agricultural Expansion: From the European steppes to Southeast Asian rainforests, once-vast ecosystems were converted to farmland, plantations, and pastoral land. Wild mammals lost both range and food sources, becoming more vulnerable to hunters and competition with livestock.

Urbanization and Infrastructure: Roads, dams, and settlements sever migration routes and isolate populations, leading to inbreeding and decreased genetic diversity.

Resource Extraction: Logging, mining, and large-scale irrigation projects altered habitats at a pace to which large mammals could not adapt.

Synergistic Impacts

In most cases, hunting alone did not drive mammals to extinction without simultaneous habitat degradation. Likewise, habitat fragmentation was often exacerbated by human-wildlife conflict, where animals forced into smaller areas would prey on livestock, prompting retaliatory or preemptive hunts. Poaching, spurred by high market values for horns, hides, or other body parts, further amplified pressures. The interplay of these factors creates "extinction vortices," where decreasing population size begets further instability through inbreeding, disease susceptibility, and reduced reproductive success.

Broader Lessons

These examples demonstrate that preventing extinction requires holistic strategies addressing multiple threats. Anti-poaching regulations, habitat restoration, community-based conservation, and ethical hunting policies must work in tandem. Moreover, global awareness and collaboration can support scientific monitoring, fund ranger patrols, and foster local stewardship. Extinction seldom emerges from a single cause; it is the culmination of intersecting social, economic, and environmental forces. Understanding this complexity can guide more effective and equitable conservation measures.

Final Thoughts on Extinct Mammals

The mammals highlighted in this chapter—Uro, Pyrenean Ibex, Javan Tiger, Caspian Tiger, Barbary Lion, Northern White Rhinoceros, Tarpan, Tasmanian Tiger, and Quagga—embody not only the beauty and diversity of life but also the fragility that arises when human activities surpass ecological thresholds. Each species carried with it a lineage shaped by millennia of evolution and played an irreplaceable role in its habitat. Their stories reveal how swiftly wildlife can disappear if our pursuits—be they for sustenance, profit, or progress—lack foresight and respect for the intricate interdependencies of nature.

From Europe's last wild cattle to Africa's vanished zebras, the overarching message remains the same: once a species is gone, no technology or breeding program can truly replicate the evolutionary tapestry that was lost. While some efforts can mimic appearances or partial genetic traits, the extinction of a species erases relationships and adaptations forged over countless generations. Our recognition of this profound loss fuels contemporary conservation efforts, aiming to ensure that future chapters of Earth's history do not repeat these tragic stories.

CHAPTER 2: EXTINCT BIRDS

Birds have long captured the human imagination in a manner that is both intimate and universal. They grace myths and legends, appear in paintings and poetry, and serve as symbols of freedom or spiritual transcendence. Yet they also navigate the same terrestrial challenges we do, often relying on habitats that are finite and fragile. Among vertebrates, birds are particularly good indicators of environmental health, their presence or absence reflecting broader shifts in ecosystems. That is why the extinction of a bird species resonates so deeply; when a creature that has taken flight since the age of the dinosaurs vanishes, it signals an alarming breakdown of the processes that maintain life on Earth. In this chapter, we turn our attention to six emblematic examples of birds that were once abundant in their respective domains but have since disappeared. Their stories range from the famous and often mythologized Dodo of Mauritius to less widely known species such as the Norfolk Island Owl, each demonstrating how hunting pressures, invasive predators, and habitat alterations can eliminate avian life. At the end of this discussion, a broader reflection on these main drivers of extinction will illustrate not only what went wrong for these birds, but how these lessons can guide modern conservation efforts.

2.1 The Dodo (Raphus cucullatus)

Of all extinct birds, few are as iconic as the Dodo, a flightless inhabitant of Mauritius in the Indian Ocean. The very word "dodo" has come to signify obsolescence, and yet the bird's actual ecological role and behavior remain elusive, filtered through the impressions of early sailors and the occasional fragmentary research that modern scientists have managed to conduct on bones and scant eyewitness accounts. Most historians and biologists agree that the Dodo was part of the Columbidae family, which includes pigeons and doves, although its exact lineage underwent significant morphological changes shaped by island isolation. Standing around a meter tall and displaying a stout, heavy body, the Dodo had atrophied wings unsuited for flight and a prominent hooked bill. Unlike many island birds, it had likely evolved to forage on fallen fruits, seeds, nuts, and possibly small crabs or shellfish along coastal areas and in the forests of Mauritius.

The story of the Dodo's extinction revolves around the sudden collision of a fauna that evolved in isolation over millennia with human activities introduced rapidly and at large scale. When Portuguese and, later, Dutch sailors arrived on Mauritius in the early 16th and 17th centuries, they found a paradise uninhabited by humans and teeming with life that had no natural fear of large predators. The Dodo, having never encountered terrestrial mammals or humans, showed little fear of the newcomers, making it dangerously easy prey. Sailors and settlers saw these birds as a convenient source of fresh meat. Mauritius also became a temporary or permanent stopover for ships engaged in the spice trade, increasing the number of ships anchoring at the island and thus the frequency of Dodo hunts. While explorers initially viewed the Dodo as somewhat unpalatable or tough to eat, hunger on long voyages often negated any culinary complaints, ensuring that the hunting continued regardless of the bird's flavor.

A second, equally devastating factor was the introduction of invasive species. Sailors and settlers unintentionally (or in some cases deliberately) brought rats, pigs, and monkeys to the island, where such mammals had never before existed. These animals, freed from the predators that might have controlled their numbers in their native habitats, proliferated rapidly. They raided Dodo nests, devouring eggs and chicks. The Dodo, having never evolved defenses against ground-based predators, and reliant on nesting strategies that were safe only in an environment free of such threats, could not adapt quickly enough to these new hazards. The clearing of forests for settlement, wood, and agriculture further encroached on the Dodo's feeding grounds, compounding the pressure. Within less than a century of European arrival, the Dodo was gone. By the late 17th century, references to living Dodos had all but ceased. A few specimens were shipped to Europe, but these were often poorly preserved, and only fragments of these original specimens remain in museums today.

Over time, the Dodo's disappearance became emblematic of human-caused extinction. Writers and naturalists turned the Dodo into a moral lesson, a symbol of the frailty of insular species subject to intense pressures from invasive predators and indiscriminate hunting. The bird's imagery appeared in works of literature, from contemporary cartoons lamenting its demise to its cameo in Lewis Carroll's "Alice's Adventures in Wonderland." Yet for all the cultural recognition the Dodo has gained, researchers still know little about its precise breeding habits, social structures, and variations across different parts of Mauritius. The limited skeletal remains offer some clues: the large beak may have assisted in cracking open tough fruits or seeds, while the robust legs allowed the bird to navigate uneven terrain. Even the bird's purportedly clumsy appearance may have been an artifact of contemporary illustrations, some of which depicted overly rotund individuals possibly based on overfed captive specimens. In the wild, the Dodo likely carried itself in a more balanced stance, consistent with a ground-dwelling pigeon adapted to a life of minimal predation threats.

The Dodo stands as the archetypal lesson in how a species supremely adapted to an isolated ecosystem can be pushed to extinction within a generation by human interference. Its story resonates in every modern discussion of invasive species control, habitat preservation, and the vulnerability of island fauna to overexploitation. While the Dodo cannot be revived, its fate underscores the responsibilities of contemporary conservation to shield the world's remaining island endemics from the same combination of hunting and invasive predation that spelled doom for this once-abundant Mauritian bird.

2.2 Passenger Pigeon (Ectopistes migratorius)

If the Dodo's story is one of isolated vulnerability, the history of the Passenger Pigeon in North America shows that even superabundance can be deceiving. During the 19th century, Passenger Pigeons existed in such staggering numbers that travelers described flocks darkening the skies for hours on end. These flocks, sometimes numbering in the billions, were among the largest aggregations of vertebrates ever recorded. They bred primarily in the deciduous forests of the eastern United States and southern Canada, subsisting on a diverse diet of nuts, seeds, fruits, and insects. Their sheer numbers were a biological marvel: the constant movement of vast flocks not only consumed vast quantities of seeds and acorns but also served as a mechanism of seed dispersal, affecting the forest composition itself. Predator avoidance was built into their enormous flocks, making it difficult for any single raptor or terrestrial hunter to pick off substantial numbers. While Native American communities had hunted Passenger Pigeons for centuries, it was primarily for local sustenance and within ecological limits that did not imperil the birds' long-term survival.

The arrival of European settlers and the expansion of commercial enterprise into North America introduced new threats that escalated well beyond subsistence hunting. With the development of railways and telegraph systems in the mid-19th century, professional pigeon hunters could quickly track large flocks, communicate sightings to fellow hunters, and ship millions of birds to urban markets. Passenger Pigeon meat, sometimes preserved in barrels or sold fresh, became a cheap source of protein in cities like New York, Boston, and Chicago. This market hunting, combined with advanced weapons, spelled calamity for a species that relied on overwhelming numbers for survival. The pigeons nested in massive colonies spanning many hectares; when found, these rookeries could be devastated in a matter of days by hunters gathering adult birds, squabs, and even eggs. Commercial hunters used nets, shotguns, and even devices that released sulfur smoke to disorient the birds. The persecution extended beyond direct killing: thousands of chicks starved when adult pigeons were slaughtered on the nest.

Deforestation of eastern North America's old-growth forests was another critical component. Passenger Pigeons depended heavily on the seasonal availability of mast—acorns, beechnuts, hickory nuts, chestnuts—which in turn thrived in extensive unbroken woodlands. As logging operations cleared these forests for timber, fuel, and farmland, the pigeons lost their primary food sources and nesting sites. Fragmented woodlands made it more difficult for such social birds to gather in large breeding colonies, weakening their reproductive capacity and compounding the effects of overhunting. Some areas saw the complete eradication of mature oaks and chestnut stands within a few decades, collapsing the once-abundant food supply.

By the early 1890s, Passenger Pigeon numbers had plummeted from billions to mere thousands. Naturalists recognized the urgent need for protective laws, but legislative actions often came too late or remained poorly enforced. The species' social and reproductive strategies, honed over millennia, were ill-suited to survival in small, scattered groups. Passenger Pigeons laid only one egg per nest, meaning their potential for population recovery was far lower than species that might lay multiple eggs or broods in a single season. As flocks dwindled, the birds also lost the protective advantage of sheer numbers, making them even more vulnerable to predators and continued hunting.

The last major nesting attempt was recorded in the Great Lakes region around the turn of the 20th century, but it produced no meaningful rebound. Martha, the last confirmed Passenger Pigeon, died in the Cincinnati Zoo on September 1, 1914. This quiet end to a species once described as "countless as the stars" remains one of the most tragic episodes in conservation history. Modern ecologists view the Passenger Pigeon's extinction as a dire warning that even species possessing colossal population sizes are not invulnerable to human greed, technological efficiency in hunting, and habitat destruction. The story of the Passenger Pigeon influenced the nascent conservation movement in the United States, spurring legislation such as the Lacey Act of 1900, which sought to regulate the interstate transportation of wildlife. Its legacy persists in contemporary debates over how to manage species that were once abundant but have experienced precipitous declines due to industrial-scale hunting or harvesting. In many ways, the Passenger Pigeon remains an instructive case in population dynamics, illustrating how a species can unravel swiftly if its breeding patterns and social structures are disrupted beyond a critical threshold.

2.3 Carolina Parakeet (Conuropsis carolinensis)

The Carolina Parakeet was the only parrot species native to what is now the continental United States, ranging from the southeastern forests to parts of the Midwest. Sporting vibrant green plumage with a bright orange forehead and yellow head, it was a striking and conspicuous inhabitant of swampy woodlands and riverine forests. Flocks of Carolina Parakeets would gather around fruiting trees, creating a spectacle of color and sound that early European settlers found mesmerizing. These parrots fed on seeds, fruits, and the seeds of cockleburs, and they often nested in large tree cavities near sources of water. Although occasionally considered agricultural pests for their fondness for orchard fruits, they also played important ecological roles by dispersing seeds throughout their range.

The decline of the Carolina Parakeet began in the 18th and 19th centuries, coinciding with significant changes in land use as European settlers transformed forested regions into farmland. Logging activities and the draining of wetlands eliminated much of the parakeet's native habitat, forcing flocks into smaller forest remnants where hunting pressures intensified. The birds' bright feathers became a commercial commodity; they were highly sought after for millinery purposes, particularly in the fashion industry for decorating women's hats. This demand fueled widespread hunting, adding another layer of pressure to already stressed populations.

Another crucial factor was the parakeet's behavioral response to hunting. Like many parrots, these birds exhibited strong social bonds and would flock together when one of their own was shot or distressed. This tendency made them easy targets for hunters, who could kill multiple birds at once when the flock circled back to investigate a downed companion. Farmers also regarded them as pests due to their predilection for orchard fruits and grain fields. In some areas, parakeet flocks were systematically exterminated to protect crops. At a time when wildlife laws were scarce or weakly enforced, the open killing of Carolina Parakeets did not evoke the moral debate that it might have triggered in a later era.

Disease may also have played a role. Some historians and biologists suspect that poultry diseases, introduced by domestic fowl, could have jumped to wild parakeet populations. Additionally, the parakeets' reliance on older, large trees for nesting cavities put them at risk as logging operations focused on the most mature and commercially valuable timber. Loss of roosting sites, combined with direct persecution, created a situation where the species had little room for rebound. Flocks that had once been counted in the thousands or tens of thousands were reduced to isolated pockets by the late 19th century. Sightings became rarer, and in some regions, the species was already considered extirpated well before the turn of the century.

By the early 20th century, small remnants of the Carolina Parakeet population persisted in Florida and perhaps parts of the lower Mississippi Valley. Unfortunately, these last stands did not survive. The precise date of extinction is not entirely clear, as unconfirmed sightings trickled in for decades. However, a commonly cited marker is the death of a captive bird named Incas at the

Cincinnati Zoo on February 21, 1918, believed to be the last of its species. The passing of the Carolina Parakeet removed from North America a charismatic seed disperser and colorful symbol of the region's once-rich avian diversity. While their memory lingers in the accounts of naturalists and the recollections of older generations, the disappearance of these parrots reflects yet another chapter in the larger narrative of how habitat destruction, direct exploitation, and ill-informed pest control measures can converge to extinguish a species.

2.4 Moho nalo (Thambetochen xanion)

Hawaiian avifauna is famous for its high rates of endemism, with numerous bird species evolving distinctive traits thanks to the archipelago's isolation. The Moho nalo (also known as the "tooth-billed goose") was one such species, an extinct genus of flightless waterfowl native to the Hawaiian Islands. The name Moho nalo roughly translates to “lost fowl” in the Hawaiian language, hinting at both its morphological uniqueness and its eventual disappearance before Western scientists could study it in the flesh. Fossil evidence suggests that Moho nalo birds were medium-sized, likely resembling a cross between a large duck and a small goose. Their bills bore unusual serrations or pseudo-teeth, adaptations that may have helped them tear apart vegetation in the lush forests of Kauai, Oahu, and other isles.

Unlike many waterfowl found on the mainland, Moho nalo species had minimal flight capability. Over generations of living in predator-free habitats, their wings atrophied and their leg structure adapted for ground foraging. This flightlessness, a recurring theme among island birds, rendered them vulnerable to the threats that arrived with the first Polynesian settlers approximately a thousand years ago. Early Hawaiians introduced domesticated animals such as pigs and dogs, cleared forest patches for agriculture, and hunted native birds for food and feathers. Flightless species like the Moho nalo, which had never developed anti-predator strategies, found themselves confronted with an onslaught of novel dangers. Their eggs and chicks were easy pickings for rats and pigs, while adult birds were hunted or forced out of their feeding grounds by deforestation and human settlements.

Archaeological digs in Hawaii have unearthed subfossil remains of Moho nalo, revealing details about their distribution, size, and potential diets. Analyzing these bones has allowed ornithologists to infer that multiple species of flightless waterfowl once occupied the islands, each likely specialized to a specific set of forested or wetland niches. The fact that so few references to Moho nalo exist in Hawaiian oral traditions suggests that they may have been driven to extinction relatively quickly after human arrival, fading out before their place in mythology or folklore could solidify in the cultural consciousness. By the time European explorers reached Hawaii in the late 18th century, the Moho nalo had been gone for centuries, leaving behind only skeletons buried in lava tubes and sinkholes.

The extinction of the Moho nalo underscores the broader fragility of island ecosystems in the face of invasive species and habitat disturbance. Hawaiian birds, from flightless waterfowl to the famous Hawaiian honeycreepers, have suffered disproportionately high rates of extinction due to introduced predators and diseases such as avian malaria. Some of the same patterns that doomed Moho nalo are still playing out for contemporary Hawaiian avifauna, indicating that lessons learned from the fossil record must inform ongoing conservation efforts. Whether one sees them as geese with teeth or ducks adapted for a forest floor, these remarkable birds reflected the evolutionary creativity found in isolated island environments, where organisms can flourish in ways unimaginable on continents. Their loss, which occurred well before Western science could catalog their habits in detail, is a reminder that extinctions can happen swiftly and quietly when the delicate balance of isolation is disrupted.

2.5 Elephant Bird (Aepyornis maximus)

The Elephant Bird, a giant flightless bird native to Madagascar, has long intrigued explorers and scholars alike. Known scientifically as Aepyornis maximus (though several related species existed), it is believed to have been one of the heaviest birds to ever walk the Earth, potentially weighing over 400 kilograms and standing up to three meters tall. In physical appearance, Elephant Birds somewhat resembled massive ostriches or emus, though they were even larger and had proportionally thicker legs. Their habitat encompassed a range of Madagascan ecosystems, from forests to subtropical woodlands, yet like other large fauna on the island, they faced the challenge of living alongside a growing human presence that arrived on Madagascar's shores around two millennia ago.

According to various accounts, the Elephant Bird laid enormous eggs, some over a meter in circumference and with a capacity equivalent to over a hundred chicken eggs. These eggs became valuable commodities for early human inhabitants, not just as a source of food but also as durable containers or vessels. There is some debate about how frequently humans hunted adult Elephant Birds, given their formidable size and strength, but eggs undoubtedly offered a simpler and more attractive resource. Overexploitation of eggs could have significantly impacted the species' reproductive success, preventing enough chicks from reaching maturity each season. Meanwhile, deforestation for agriculture, combined with slash-and-burn techniques, progressively altered Madagascar's landscapes. In addition to direct exploitation, the introduction of domesticated animals and later invasive species may have stressed the Elephant Bird's foraging grounds and contributed to nest predation.

The stories that swirl around the Elephant Bird often mingle with legend. Some scholars speculate that tales of giant birds like the roc in the Arabian Nights were partly inspired by travelers' accounts of Aepyornis eggs or skeletal remains. Marco Polo and other medieval chroniclers hinted at giant birds in the Indian Ocean region that could carry off elephants, though these were likely exaggerations or amalgams of real sightings and mythic lore. Regardless, the Elephant Bird's actual existence was confirmed by scientific study of subfossil bones and eggs beginning in the 19th century. Today, museum collections around the world include reconstructed skeletons and intact eggs, but these relics can only hint at the bird's behavior, ecology, and significance within Madagascar's pre-human or early-human ecosystems.

The precise timing of the Elephant Bird's extinction remains somewhat debated. Some researchers believe that they may have survived in remote regions of Madagascar until as recently as five or six centuries ago, while others argue for an earlier disappearance. In any case, the synergy of habitat changes, hunting, and egg collection likely spelled doom for these flightless giants well before modern ornithology could observe them alive. Modern parallels can be seen in other large, flightless birds and mammals across the globe, highlighting the vulnerability of species that reproduce slowly and depend on stable habitats for foraging and nesting. The Elephant Bird's massive eggs represent both a scientific curiosity and a poignant metaphor for how the largest forms of life can be undone by seemingly small but consistent human pressures. As a key piece of Madagascar's lost megafauna, it reminds us that the island's current biodiversity crisis began long ago and may accelerate if contemporary threats such as logging, mining, and climate change are not managed sustainably.

Main Causes: Hunting, Invasive Species Introduction

Across these six cases—the Dodo, Passenger Pigeon, Carolina Parakeet, Moho nalo, Elephant Bird, and Norfolk Island Owl—the recurring factors of hunting and invasive species introduction emerge as the major drivers of extinction. Yet the nuances vary. For the Dodo and Elephant Bird, the removal of eggs from nests was as devastating as the hunting of adult birds, illustrating how vulnerable breeding cycles can be when humans or invasive mammals exploit them. For the Passenger Pigeon, commercialized market hunting amplified the already significant pressures of habitat loss, turning what were once unthinkably large flocks into scattered remnants that could not recover. For the Carolina Parakeet, direct extermination as a perceived agricultural pest merged with the millinery trade's desire for brightly colored feathers, a dual blow that ravaged populations already suffering from deforestation.

Invasive species introduced by human agency represent an equally pervasive threat, especially on islands. The story of the Dodo proves how a single island's isolation can be shattered by rats and pigs arriving on ships. The Norfolk Island Owl, meanwhile, encountered a web of competition and predation by cats, rats, and other newcomers that it had not evolved to withstand. Hawaiian flightless waterfowl such as the Moho nalo similarly encountered dogs, pigs, and rats that turned their once-safe nesting grounds into veritable death traps. The complex interplay of direct exploitation and biological invasions underscores the fragility of birds that are either adapted to highly specialized ecological niches or reliant on large-scale phenomena like mass nesting. Once these patterns are disrupted, whether by introduced predators or relentless hunting, the potential for recovery diminishes rapidly.

In examining these examples, it is clear that extinction can occur with startling swiftness when the evolutionary adaptations that allowed a species to flourish in its natural environment clash with external pressures that arrive suddenly. The reasons behind that clash—be they commercial hunting networks, invasive predators, or large-scale deforestation—often trace back to human economic, social, or cultural practices. The Passenger Pigeon and Carolina Parakeet reveal how the industrialization of hunting and the commodification of wildlife resources in rapidly growing societies can decimate populations that are poorly protected by legislation. The island-based extinctions, from the Dodo to the Moho nalo, illuminate just how lethal the introduction of non-native species can be when the ecosystem is poorly equipped to adapt to new predatory or competitive threats.

Ultimately, these histories inform present-day conservation. They highlight the urgency of mitigating invasive species, the importance of preserving sufficient habitat to maintain breeding colonies, and the necessity of regulating hunting so that it does not devolve into unchecked exploitation. They also underscore the ethical dimension of our relationship to wildlife: species like the Carolina Parakeet or the Passenger Pigeon were destroyed not out of dire necessity, but often from shortsighted economic motives, fashion trends, or simple ignorance of the consequences. It is a reminder that biodiversity, while resilient in many ways, has limits beyond which recovery is impossible.

For modern ornithologists, these stories serve as cautionary tales and motivate the study of how remaining bird species might be protected from parallel fates. Conservation tools such as habitat restoration, invasive predator eradication, and enforced legal protections can sometimes stabilize or revive endangered populations if implemented in a timely manner. Yet the cautionary lesson from these extinct birds is that the window of opportunity is finite. The Dodo and Elephant Bird likely vanished before their ecosystems could be fully documented, cutting off generations of researchers from unraveling the intricate web of relationships those birds once participated in. The Passenger Pigeon and Carolina Parakeet, once ubiquitous across vast swathes of North America, disappeared in plain sight, prompting social reflection on how unstoppable modern commerce and expansion can be if left unchecked. The Norfolk Island Owl, overshadowed by more famous extinctions, still offers a sobering vision of how an isolated predator can quietly decline and then vanish amid habitat disruption and unchecked invasives.

In every case, hunting and invasive species are intertwined with other threats, including habitat loss, disease, and shifts in local climate. But the underlying pattern is consistent: the actions of humans, deliberate or otherwise, introduced pressures these birds could not withstand. That consistency is what makes these extinctions so haunting and so instructive. They are not anomalies in the grand sweep of natural history; they are emblematic outcomes of processes that continue to shape the fate of countless species worldwide. By understanding why these birds failed to survive, conservation science can better adapt management strategies to prevent other avian tragedies. When we look back on the fates of the Dodo, the Passenger Pigeon, the Carolina Parakeet, the Moho nalo, the Elephant Bird, and the Norfolk Island Owl, we see a tapestry of losses that speaks less to the inevitability of extinction and more to the urgent need for vigilance, regulation, respect for ecological limits, and a willingness to learn from past mistakes. These extinct birds remind us that even though flight may confer a powerful evolutionary advantage, no wings can outrun the combined forces of unchecked hunting and the rampant introduction of invasive species.

CHAPTER 3: EXTINCT REPTILES

Reptiles, a group that has inhabited Earth for hundreds of millions of years, have adapted to remarkably diverse environments—from the highlands of oceanic islands to the arid expanses of continental deserts. Yet, even these resilient survivors of past mass extinctions are far from immune to the intense pressures of the modern era. In this chapter, we examine four emblematic examples of reptile extinctions: the Pinta Island Tortoise, the La Palma Giant Lizard, the Seychelles Giant Tortoises, and the Santo Stefano Island Lizard. Each case reveals a blend of habitat destruction, introduced predators or competitors, and human exploitation, culminating in the collapse of species that had once thrived in isolated niches. A final discussion ties these stories together, highlighting the overarching factors—particularly habitat degradation and invasive species—that have propelled reptile populations to the brink and, in some instances, beyond.

3.1 Pinta Island Tortoise (Chelonoidis abingdonii)

When Charles Darwin first visited the Galápagos Archipelago in 1835, he observed that each island hosted its own distinct race or species of giant tortoise, a phenomenon that contributed to his thinking on natural selection and adaptive radiation. Among these remarkable tortoises, the Pinta Island Tortoise (Chelonoidis abingdonii) would become an infamous symbol of extinction in modern times. Pinta Island, a small and relatively remote part of the Galápagos chain, once harbored a population of giant tortoises characterized by their saddleback shells—an adaptation that allowed them to reach higher for vegetation in an otherwise sparse environment. The archipelago's arid zones and patchy vegetation made the tortoises essential seed dispersers, as they consumed cacti, leaves, and fruit, distributing viable seeds across the island's volcanic terrain.

Centuries before any serious conservation efforts took root, visiting ships introduced goats and other non-native animals to Pinta Island. Sailors often released goats or pigs intentionally, hoping to create a source of fresh meat during future voyages, but these introduced species proliferated uncontrollably in the absence of their natural predators. The goats, in particular, ravaged the vegetation upon which the native tortoises depended, stripping hillsides of shrubs, grasses, and cacti. The fragile ecological balance that had sustained the tortoises for millennia collapsed, as the tortoises struggled to find enough food in the increasingly barren landscape. In addition to competition for resources, direct hunting also took a toll. Seafarers viewed giant tortoises as a convenient larder, since they could be kept alive on ships for months without food or water, providing fresh meat throughout a voyage.

By the mid-20th century, biologists realized that the Pinta Island Tortoise was on the brink of extinction. Efforts to remove feral goats from the island came too late to save most of the tortoises. In 1971, scientists searching Pinta discovered a single male tortoise living in isolation. Dubbed "Lonesome George," this solitary reptile became a poignant reminder of irreversible biodiversity loss. George was moved to the Charles Darwin Research Station on Santa Cruz Island, where conservationists attempted to breed him with females of closely related Galápagos tortoise species, hoping that some fraction of Pinta's genetic legacy might persist. Despite decades of these efforts, no viable offspring resulted, and Lonesome George died on June 24, 2012, marking the extinction of Chelonoidis abingdonii as a distinct taxon.

George's passing generated global headlines, prompting renewed reflection on the complexities of invasive species control, habitat restoration, and the importance of acting before populations reach a single remaining individual. In the years following his discovery, the Galápagos National Park intensified efforts to eradicate goats from Pinta Island and neighboring islands. Remote hunting campaigns, sometimes involving helicopters, succeeded in removing large numbers of feral goats, allowing vegetation to recover and setting the stage for possible ecological restoration. Yet, with no surviving Pinta Tortoises, the island's original giant herbivores and seed dispersers remain absent, likely altering plant community dynamics in ways that may never fully revert to the pre-invasion state.

Although the Pinta Island Tortoise is now extinct in a strict biological sense, some conservationists have continued searching for hybrid tortoises on other islands where Pinta individuals might have interbred long ago. Genetic sampling of tortoises in remote volcanic regions has yielded evidence of partial Pinta ancestry, though no purebred specimens have been discovered. These findings hint at a glimmer of hope for future breeding programs that might resurrect something approaching the Pinta lineage if a suitable pool of hybrid genetics can be harnessed. However, even if such a project were to succeed scientifically, the cultural and scientific meaning embodied by Lonesome George underscores that a reconstituted genotype is not the same as preserving the unbroken lineage that once defined Pinta Island's ecology.

The demise of the Pinta Island Tortoise resonates far beyond the Galápagos, illustrating how easily isolated reptiles can be outcompeted and pushed toward oblivion once alien species disrupt longstanding ecological balances. Humans can be both the destroyers and potential saviors of such unique life forms. In the case of Pinta, recognition of the tortoise's plight came too late to rescue its pure genetic identity, but it contributed significantly to stricter conservation measures within the archipelago. In the broader realm of wildlife preservation, George's story continues to be evoked as a reminder that last-minute attempts at captive breeding often fail when the underlying habitat disturbances remain unaddressed. This narrative of what was lost—and what might still be recovered—serves as a guiding example in the ongoing struggle to protect other vulnerable reptiles in similarly fragile island ecosystems.

3.2 Seychelles Giant Tortoise (Cylindraspis spp.)

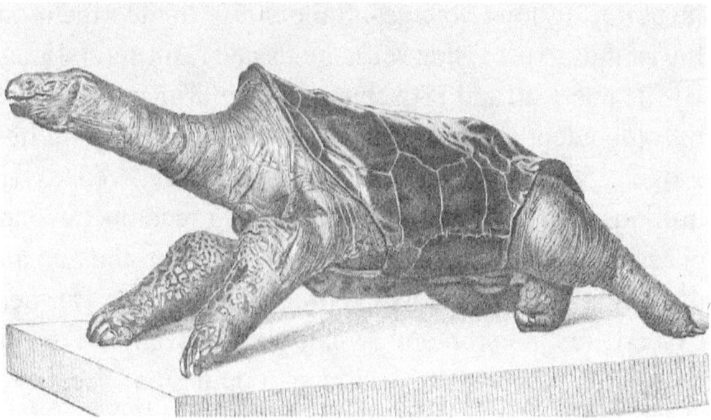

Scattered across the western Indian Ocean, the Seychelles once hosted an assortment of reptiles that reflected the region's status as a biodiversity hotspot. Large, herbivorous tortoises were among the most noteworthy inhabitants of these island groups, offering a living analogue to the more famous Galápagos and Aldabra tortoises. Historical records indicate that multiple species and subspecies of giant tortoises once thrived on various islands in the western Indian Ocean, often grouped under the genus Cylindraspis. Although the Aldabra Giant Tortoise (Aldabrachelys gigantea) remains extant on Aldabra Atoll, others in neighboring archipelagos, including the Mascarenes and some parts of the Seychelles, disappeared before modern science could detail their biology extensively.

In the early Age of Exploration, navigators sailing around the Cape of Good Hope or across the Indian Ocean discovered these island tortoise populations and recognized them as a valuable food resource. Tortoises, much like their Galápagos counterparts, could be kept alive in ship holds, providing a steady supply of fresh meat during long voyages. Sailors also hunted them on the islands to provision settlements and refitting stations. By the late 17th and 18th centuries, seafaring nations had become adept at exploiting the region's island wildlife. Large tortoise populations that had evolved over millions of years, often with no native predators beyond occasional birds of prey, could not withstand the human onslaught. Their slow reproductive rates further exacerbated the problem; female tortoises laid relatively few eggs each year, making population recovery painfully slow if adult mortality rates soared.

Compounding direct exploitation was the introduction of pigs, rats, and other feral animals to these islands. Tortoise eggs and hatchlings became easy meals for these newcomers, disrupting the species' life cycle and recruitment. Habitat changes also played a significant role. Sailors and settlers cleared coastal forests and manipulated freshwater sources, which might have served as essential watering holes or breeding areas for tortoises. Over time, entire island populations of Cylindraspis species vanished, sometimes documented only through sailors' notes describing their abundance mere decades earlier.

A salient example is the now-extinct Seychelles Giant Tortoise that formerly inhabited several smaller islets within the archipelago, distinct from the Aldabra Tortoise but similarly massive and long-lived. By the time naturalists of the 19th and early 20th centuries attempted to study these tortoises systematically, authentic wild populations had mostly disappeared. A few individuals may have been kept in captivity, but consistent breeding was never established. As the numbers fell, morphological and genetic distinctions became blurred, leading to confusion about precisely which species or subspecies inhabited which islands. The repeated pattern of extinction on scattered islands left behind incomplete records and a taxonomic puzzle that herpetologists and paleontologists are still trying to piece together using subfossil remains, museum specimens, and sporadic historical accounts.

Debate occasionally arises over whether any pure Seychelles Giant Tortoises survive, concealed among captive populations labeled as Aldabra tortoises or other related forms. Genetic analyses have, in some instances, uncovered traces of distinct lineages, suggesting that hybridization might have occurred among the remaining tortoises after multiple introductions to captivity. Yet, the consensus is that the original, wild forms of many Seychelles and Mascarene Giant Tortoises (formerly included in the genus Cylindraspis) are extinct. Some conservationists hold out hope that selective breeding of hybrids or back-breeding from living relatives could, in theory, restore at least some morphological and ecological characteristics of these lost lineages. However, the conservation world is divided on the ethics and efficacy of devoting significant resources to reviving a tortoise population that would necessarily differ in subtle ways from the extinct forms.

The story of the Seychelles Giant Tortoises echoes that of the Pinta Island Tortoise, but on a broader scale. Multiple island forms fell victim to similar pressures—unregulated hunting, invasive predation, and habitat alteration—over a remarkably short span of history. This repeated tragedy underscores how vulnerable insular reptile species can be when confronted with colonial exploitation. The legacy of the Cylindraspis extinctions lingers as modern conservationists strive to protect the Aldabra Giant Tortoise and other surviving Indian Ocean reptiles from a similar fate. Restoration efforts for smaller islands in the Seychelles often involve removing invasive mammals, stabilizing soil erosion, and allowing vegetation to recover, thus creating conditions under which reintroduced or remnant tortoises might thrive. While these endeavors cannot bring back the original Seychelles Giant Tortoise in its pure evolutionary form, they may provide a critical hedge against further reptile extinctions in an island realm that once teemed with giant shelled herbivores.

3.3 Santo Stefano Island Lizard (Macroscincus coctei)

Situated off the coast of Africa, the Cape Verde archipelago spans a chain of volcanic islands that harbor numerous endemic plants and animals. Among these was the Santo Stefano Island Lizard, scientifically named Macroscincus coctei, which occupied a unique ecological niche in an environment shaped by high temperatures, erratic rainfall, and steep volcanic slopes.

Also referred to as Cocteau's Skink, this reptile was an unusually large member of the skink family, with a robust body and short limbs adapted for burrowing or navigating rocky crevices. Historical accounts and subfossil evidence suggest it favored dry, rocky areas with sparse vegetation, feeding on insects, small vertebrates, and perhaps plant material, making it an omnivore well suited to the archipelago's challenging environment.

The early explorers and settlers who arrived in Cape Verde introduced goats, cats, and other domesticated animals to the islands in a bid to create sustainable food supplies. These non-native species quickly became feral in some areas, transforming fragile ecosystems that had evolved without large mammalian grazers or predators. For the Santo Stefano Island Lizard, cats posed a particular danger, preying on both juveniles and adults. The presence of goats also altered the landscape by heavily browsing on shrubs and grasses, hastening erosion and reducing ground-level cover that skinks might have used for shelter. In this open, degraded environment, the lizards became more exposed to predators and struggled to find refuge.

Sporadic reports indicate that humans also hunted Macroscincus coctei for food, although it may not have been a primary target. Local islanders and passing sailors occasionally captured these large skinks, perhaps finding them easy prey given their size and terrestrial habits. The exact timeline of the lizard's decline is uncertain, but by the late 19th or early 20th century, naturalists seeking specimens on the islet of Santa Luzia or other small Cape

Verdean islands discovered that the species had become exceedingly rare or perhaps already extinct. Poor record-keeping and the remoteness of these islands make it difficult to pinpoint the final date of disappearance, but most experts agree that Macroscincus coctei no longer survives in the wild.

The absence of extensive contemporary observations hampers our understanding of how this skink interacted with the Cape Verde environment. Without direct ecological studies, scientists can only infer from morphological traits and the distribution of subfossil remains that the Santo Stefano Island Lizard played a functional role as both predator and seed disperser. Some large skinks elsewhere in the world consume fruits and inadvertently spread seeds through their droppings, thus influencing plant community dynamics. The disappearance of Macroscincus coctei may have led to subtle but significant changes in the local vegetation and in the population dynamics of insects or small vertebrates that once found themselves on the skink's menu.

In recent years, Cape Verdean authorities and international conservation groups have taken steps to protect and restore critical habitats on certain islands within the archipelago. Measures include the control or eradication of feral cats and goats, as well as the establishment of nature reserves. Yet these actions commenced long after the likely extinction of Macroscincus coctei. A handful of expeditions and searches have been launched in remote parts of the archipelago, hoping to stumble upon a surviving pocket of these skinks, but none have yielded conclusive evidence. Much like the search for the La Palma Giant Lizard in the Canary Islands, these efforts reflect a broader yearning within the conservation community to discover that a presumed-extinct reptile might still endure in overlooked refuges.

Macroscincus coctei's fate underscores the ease with which a specialized island reptile can be driven to extinction when confronted by introduced predators and widespread habitat degradation. Although the species' morphological uniqueness might have piqued curiosity among collectors and explorers, no formal conservation plan was implemented while the lizard still had a chance to recover. Similar stories resonate across island chains worldwide, where endemic reptiles and amphibians vanish quietly due to a combination of direct exploitation, feral predators, and habitat loss. The Santo Stefano Island Lizard now endures only in a handful of preserved museum specimens, a tangible reminder of how quickly evolution's most distinctive creations can slip through our fingers when protective measures lag behind the speed of ecological disruption.

3.4 Main Causes: Habitat Destruction and Invasive Species

The extinction of reptiles such as the Pinta Island Tortoise, the La Palma Giant Lizard, the Seychelles Giant Tortoises, and the Santo Stefano Island Lizard reveals two dominant themes that surface again and again in island ecosystems: habitat destruction and the introduction of invasive species. Although each reptile inhabited a different part of the globe, from the Galápagos to the Canary Islands, the underlying pressures that led to their downfall share striking parallels. There is no single factor that can be isolated in each case, yet examining how these forces interact provides a clearer sense of why reptiles—particularly those on islands—are so vulnerable.

Habitat destruction, whether through deforestation, overgrazing by introduced livestock, or large-scale agricultural conversion, remains a pervasive threat. In island contexts, the scale of these habitat changes is often compressed. A relatively small amount of agricultural expansion or forest clearing can eliminate a significant fraction of a species' range. Pinta Island was overrun by goats that devoured the vegetation needed by the tortoises. In the Cape Verde archipelago, goats and other livestock similarly reduced plant cover, exposing the Santo Stefano Island Lizard to predators and harsh environmental conditions. In the Canary Islands, increased human settlement and agricultural activities took over the prime habitats that large lizards depended on, limiting their ability to find food and secure nesting sites. Habitat modifications tend to have compound effects over time: once an area loses its native vegetation, erosion can set in, further degrading the environment and making it even less hospitable for species that have spent millennia adapting to specific ecological niches.

Simultaneously, invasive species have proven devastating for island reptiles, which often evolved in the absence of mammalian predators. The introduction of cats, rats, pigs, or goats radically alters the food web. Cats are agile hunters that decimate lizard populations, tortoise hatchlings, and other native fauna. Rats, in turn, are prolific breeders that prey on eggs and young reptiles, while also consuming native plants and seeds, thereby compounding habitat stress. Pigs and goats do not necessarily target reptiles directly, but they degrade the environment or compete for resources, thereby undermining the fragile equilibrium that sustained these unique reptiles for so long. The historical accounts of explorers in the Indian Ocean, who systematically harvested giant tortoises and introduced feral animals, reveal how swiftly a seemingly abundant population can plummet if breeding cycles are disrupted and adult mortality spikes.

Another layer in these stories is the role of direct human exploitation, which often goes hand in hand with habitat destruction and invasive species. Giant tortoises were prized food sources for maritime ventures, prized precisely because they were easy to capture and keep alive on ships for extended periods. Large lizards, although less commonly targeted as a staple, were at times hunted for meat or collected for curiosity. Even modest levels of exploitation can prove lethal for reptiles that reproduce slowly, lay relatively few eggs, or depend on stable adult populations for species continuity. A single ill-timed wave of hunting or collection can tip an already declining population into an irreversible trajectory.

Fragmentation of populations also emerges as a repeated element. As habitats become divided by farmland, roads, or degraded patches, remaining reptiles are forced into smaller, isolated pockets. Genetic exchange declines, and the risk of localized catastrophe—from a cyclone to a disease outbreak—increases. In the case of the Seychelles Giant Tortoises, scattered island populations vanished one by one. Similarly, the last known Pinta Island Tortoise lived alone, highlighting the grim consequences of fragmentation when a species is reduced to a single, non-breeding individual.

From a conservation standpoint, these extinct reptiles illustrate the urgent need to address habitat preservation, invasive species management, and responsible stewardship long before a species is reduced to a handful of survivors. Eradicating or controlling goats, rats, and cats on islands has proven possible in multiple high-profile successes, yet such operations demand careful planning, funding, and societal support. Programs to restore vegetation can also be effective, as has been demonstrated in parts of the Galápagos, where removing invasive herbivores has allowed native flora to rebound. The question remains whether these efforts can be enacted fast enough or broadly enough to prevent further reptile extinctions. In many cases, by the time conservation steps are taken, species have already declined so severely that genetic bottlenecks and demographic instabilities hinder any chance of robust recovery.

Looking ahead, the cautionary tales from these reptiles point toward a broader principle: islands function as microcosms for global environmental challenges. The processes that drive extinctions on islands—habitat destruction, invasive species, and unsustainable exploitation—are mirrored on continents, albeit on a larger scale. The difference is that island species typically have no "elsewhere" to go and no recolonization pathways if local extirpations occur. The line between survival and extinction is perilously thin when an entire species is confined to a single island or a few closely grouped islets. Nevertheless, these cases also demonstrate that human-induced threats, though immense, are not necessarily unstoppable. If societies muster the will to control invasive species, curb destructive practices, and rehabilitate degraded habitats, it is possible to stabilize or even revive certain reptile populations before they cross the threshold of irreversibility.

The extinct reptiles discussed in this chapter are thus not mere curiosities of the past; they offer critical lessons for contemporary conservation efforts. Whether one mourns the solitary passing of Lonesome George or laments the quiet disappearance of the La Palma Giant Lizard, the central message remains consistent: the arc of extinction can be swift once invasive predators, habitat fragmentation, and overexploitation take hold. These stories should galvanize ongoing and future projects aimed at protecting extant island reptiles. They should also serve as an impetus for more robust global commitments to preventing the spread of invasive species and addressing the habitat needs of endangered wildlife. In doing so, we might ensure that the tragic fates of the Pinta Island Tortoise, the La Palma Giant

Lizard, the Seychelles Giant Tortoises, and the Santo Stefano Island Lizard will stand not only as epitaphs for what has been lost, but also as motivation to safeguard what still remains.

CHAPTER 4: EXTINCT FISH

Fish are among the earliest vertebrates to have evolved on Earth, and for over 500 million years, they have populated virtually every aquatic habitat—from deep ocean trenches to high mountain lakes. Their diversity, adaptability, and ecological importance are hard to overstate. Fish provide sustenance for countless species, shape aquatic food webs, and even influence global biogeochemical cycles through nutrient transport. Yet as human demand for water resources grows and fishing technologies become more efficient, fish populations have come under significant pressure. Entire species can vanish when pollution degrades their habitats or when intense harvesting depletes their numbers beyond recovery. This chapter examines five fish species that have been driven to extinction in modern times: the Tecopa Pupfish, the Lake Geneva Whitefish, the Lake Ontario Salmon, the Gravenche, and the Potosi Cichlid. Through these examples, we see how pollution, overfishing, habitat fragmentation, and other anthropogenic factors interact to erase lineages that once teemed in rivers and lakes. We conclude by reflecting on the main causes—pollution and overfishing—that emerged repeatedly as the catalysts for these extinctions, and what these tragedies imply for the future of aquatic biodiversity worldwide.

5.1 Tecopa Pupfish (Cyprinodon nevadensis calidae)

The Tecopa Pupfish was native to the hot springs outflows in the Tecopa area of California's Mojave Desert. Pupfish in general (genus *Cyprinodon*) represent a remarkable evolutionary success story, having adapted to some of the most extreme freshwater environments in North America. These small, hardy fish often survive in habitats with high salinity or temperature fluctuations that would kill most other species. The Tecopa Pupfish was no exception. For thousands of years, it thrived in the warm waters that emerged from the ground at Tecopa Hot Springs, an environment where water temperatures could exceed 40°C (104°F). Despite these inhospitable conditions for many other organisms, the Tecopa Pupfish found a niche, feeding on algae, tiny invertebrates, and detritus while completing its life cycle in shallow outflows just a few centimeters deep.

The downfall of this species began in the mid-20th century, when development interests targeted the natural hot springs for tourism. Entrepreneurs saw the potential to build bathhouses, hotels, and recreational facilities around the springs, capitalizing on the region's supposed therapeutic qualities and picturesque desert landscapes. To make the springs more appealing for visitors, local developers channeled or altered the water flow, sometimes cooling the outflows or diverting them entirely. These modifications drastically changed the microhabitats where the Tecopa Pupfish reproduced and fed. Pupfish depend on consistent water temperature regimes; even small alterations in flow patterns can disrupt breeding cycles. By channelizing the outflows, the temperature gradients that had once allowed the fish to occupy specific sections of the stream were flattened, leaving them without the finely tuned hot-cool transitions they required to thrive.

Tourism-related development also led to increased sedimentation and pollution. Soaps, detergents, and other chemicals introduced by bathers and resort facilities entered the pupfish's habitat. Over time, these contaminants settled into the shallow waters, affecting oxygen levels and water chemistry. Unlike some pupfish that can tolerate extreme salinity or temperature, the Tecopa Pupfish was not well-adapted to chemical pollutants or silt-laden water. The ecological balance that had persisted for centuries in the desert's harsh conditions quickly deteriorated under these new anthropogenic pressures.

By the 1960s, biologists had begun sounding alarms over the declining pupfish populations. Conservation measures like regulated water use or habitat restoration were proposed but implemented inadequately. Local economic interests often clashed with environmental concerns, leading to partial or delayed reforms. As populations dwindled, scientists scrambled to collect specimens for study or potential captive breeding, but pupfish are notoriously sensitive to changes in temperature, water chemistry, and flow rates. Captive populations failed to establish a sustainable foothold, and wild numbers continued to crash. By the late 1970s, the Tecopa Pupfish was functionally extinct, and it was officially declared extinct in 1981. Its demise made it one of the first fish species in the United States to be recognized as lost due to direct human actions in a relatively recent historical period.

The extinction of the Tecopa Pupfish highlights a paradox: although these fish evolved to endure high temperatures and low oxygen levels, they could not survive even moderate habitat alterations caused by human development. It also underscores the vulnerability of springs and desert wetlands, which often appear robust because of their extreme conditions but are, in fact, delicately balanced ecosystems. Pupfish generally serve as an indicator of the health of desert aquatic habitats; losing them means an entire web of smaller invertebrates, algae, and microorganisms is also in jeopardy. Had stricter habitat protections or more ecologically sound development plans been in place, the Tecopa Pupfish might have persisted in its steaming desert refuges. Instead, its final disappearance stands as a cautionary tale about the swift ecological consequences of poorly managed tourism and recreational expansion.

5.2 Lake Geneva Whitefish (Coregonus fera)

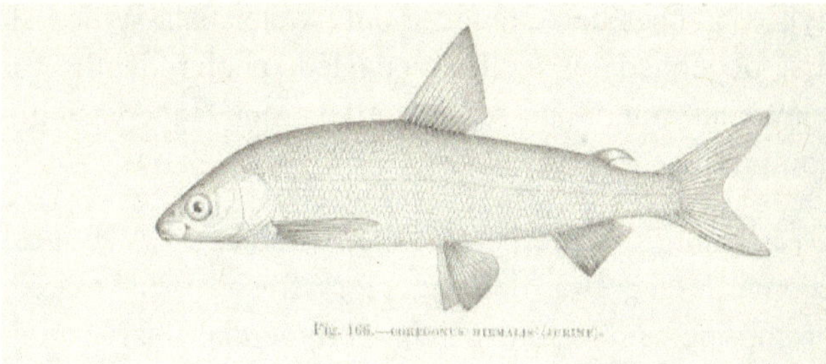

The Lake Geneva Whitefish, *Coregonus fera*, was once a staple resource for communities along the shores of Lake Geneva (also known as Lac Léman) in Switzerland and France. The genus *Coregonus* includes various whitefish species that inhabit cold and often deep northern waters; some have historically supported major fisheries in Europe and North America. For centuries, local fishermen relied on *Coregonus fera* and other whitefish for commerce and subsistence. The fish were prized not only for their delicate flavor but also for their relatively predictable spawning runs and schooling behavior, which made them accessible targets for small-scale, sustainable fisheries.

However, as industrialization progressed in the 19th and 20th centuries, Lake Geneva experienced a surge in pollution levels. Rapid urbanization around Geneva, Lausanne, and other lakeside towns led to a spike in nutrient runoff from sewage, agriculture, and industry. Fertilizers, detergents, and other pollutants flowed into the lake, causing eutrophication—an over-enrichment of nutrients that stimulated algal blooms. These blooms, in turn, robbed the deeper layers of the lake of dissolved oxygen when algae died and decomposed, creating hypoxic zones unsuitable for cold-water fish. Whitefish that depended on well-oxygenated, cooler depths for spawning found their reproductive grounds increasingly compromised.

Overfishing exacerbated the problem. As local economies grew, demand for fresh fish soared, prompting more intensive commercial harvesting. Innovations in fishing gear, such as more efficient gillnets and motorized boats, allowed fishers to capture whitefish at higher rates, including during spawning aggregations. While historical records suggest that local fisheries were once relatively controlled by tradition and communal agreements, the pressures of modern markets undermined these customs. Coupled with a lack of robust regulatory frameworks during the early and mid-20th century, the whitefish populations were unable to replenish themselves in the face of constant extraction.

Conservationists and fisheries biologists tried various remedial measures, including stocking Lake Geneva with hatchery-bred coregonid fry in the hope of bolstering declining populations. Unfortunately, hatchery programs often rely on broodstock that may not match the specific genetic lineage of the wild *Coregonus fera*, introducing potential hybridization problems or simply failing to thrive. Additionally, the fundamental issue of eutrophication persisted well into the late 20th century, as local authorities struggled to upgrade sewage treatment plants and limit agricultural runoff. Even when pollution controls started to improve in the 1970s and 1980s, the whitefish's decline had already reached a critical point.

By the late 20th century, scientists recognized that *Coregonus fera* had effectively disappeared from the lake. While a few other coregonid species in Lake Geneva survived or were maintained via stocking programs, the original lineage known as the Lake Geneva Whitefish could no longer be found. Historical accounts that once described abundant catches dwindled to anecdotes about a fish that had "vanished without a trace." Some taxonomists argue that remnants of *Coregonus fera* genes may persist in hybrids with other introduced or stocked whitefish, but the distinct population that formed a crucial cultural and economic resource for lakeside communities is gone.

The disappearance of the Lake Geneva Whitefish serves as a potent illustration of how pollution and overfishing can interact to drive an aquatic species to extinction. Had either pressure been managed effectively, the fish might have remained resilient. If water quality had been safeguarded, the fish could have withstood moderate harvesting. Conversely, if fishing had been limited, perhaps the species could have adapted to changing water conditions until pollution controls were implemented. Instead, the lethal combination of eutrophication and unsustainable harvesting left no margin for recovery. Today, Lake Geneva's improved water quality stands as a testament to successful pollution control measures, but it also reminds locals that reversing ecosystem damage does not always bring back species that have already slipped away.

5.3 Gravenche (Coregonus hiemalis)

In the alpine lake systems of Europe, multiple coregonid fish once held ecological and economic significance. The Gravenche, *Coregonus hiemalis*, was closely related to the Lake Geneva Whitefish but populated other nearby lakes in the Rhône basin and possibly overlapping zones. Like its close relatives, Gravenche preferred the well-oxygenated deep waters, migrating seasonally to shallower areas for spawning. Throughout much of the 19th and early 20th centuries, local fishermen harvested Gravenche along with other whitefish, selling the catch in local markets and restaurants. Whitefish were a cornerstone of the region's culinary traditions, with recipes and festivals centered around their seasonal abundance.

By the mid-20th century, however, the same suite of pressures—eutrophication, pollution, and intensive fishing—took a toll on *Coregonus hiemalis*. The rapid industrialization of Switzerland, France, and neighboring regions heightened nutrient inflow into alpine lakes. Agricultural expansion along mountain valleys contributed further to fertilizer runoff, spurring algal blooms and subsequent oxygen depletion. Meanwhile, improved fishing techniques enabled local fishers to remove large numbers of adult Gravenche even in deeper offshore waters. The fish's relatively slow reproductive rate and sensitivity to oxygen-poor conditions during spawning left little room for error. Once the population began declining, it could not rapidly rebound.

Adding to these challenges was the competition and genetic swamping introduced by stocking programs. As in Lake Geneva, fishery authorities sometimes released other whitefish species or strains to boost commercial yields. While these efforts sought to mitigate economic losses, they often confused the native population's breeding cycles. Hybridization might have occurred, diluting the distinct lineage of Gravenche. The subtle morphological and genetic differences that had evolved over centuries to suit specific alpine lake environments became muddled. Some experts believe that even before *Coregonus hiemalis* was formally recognized as extinct, it may have been functionally gone, persisting only as hybrids that eventually merged into introduced whitefish stocks.

Throughout the 1970s and 1980s, repeated surveys found fewer and fewer Gravenche. By the time water quality improvements began to yield benefits, it was too late for the original populations. The last confirmed captures occurred sporadically, sometimes only a handful of fish per fishing season, suggesting that pockets of survivors held on but were not reproducing at levels sufficient to sustain the species. Eventually, these remnant groups disappeared. The loss of the Gravenche and related alpine whitefish highlights a pattern of multiple extinctions in European lake systems. As water pollution controls tightened in the latter part of the 20th century, it became clear that some species had crossed ecological thresholds from which they could not return.

The Gravenche's fate, like that of the Lake Geneva Whitefish, demonstrates the vulnerability of lacustrine (lake-dwelling) fish to even moderate but sustained anthropogenic changes. Alpine lakes might appear pristine, but they are actually closed systems with limited capacity to flush out pollutants. When nutrient levels spike, oxygen levels drop in deep waters, compromising spawning success for cold-water fish. Overfishing can exacerbate these conditions by removing breeding adults faster than populations can rebound. Stocking introduced species or hybrids, rather than controlling the root problems, can inadvertently doom native forms by homogenizing gene pools. For local communities that once prided themselves on the distinct flavors and traditions tied to Gravenche, its disappearance is a cultural as well as ecological loss. Today, tourists visiting the region can still dine on "whitefish," but it is rarely, if ever, the authentic local species that once defined alpine lake fisheries.

5.4 Potosi Cichlid (Herichthys minckleyi)

Shifting from cold-water habitats to a desert oasis, the Potosi Cichlid inhabited a set of springs and small lakes in the region around Potosí in northeastern Mexico. These unique aquatic systems, often referred to as "cuatrociénegas" (though geographically distinct from the more famous Cuatro Ciénegas region in Coahuila), were fed by natural springs and characterized by clear, mineral-rich waters. The cichlids of Mexico's northern desert region are renowned for their rapid speciation and remarkable adaptations. Many cichlids display specialized feeding structures, elaborate reproductive behaviors, and distinct coloration that reflect their isolation in separate springs or sinkholes over evolutionary timescales.

The Potosi Cichlid depended on stable spring flows and minimal disturbance to the local hydrology. However, starting in the mid-20th century, the region experienced intense agricultural development. Farmers sought to expand crop fields and pastureland, necessitating irrigation projects that tapped the natural springs. Water was diverted or over-extracted to feed canals, lowering water levels in the cichlid's habitat. As these desert springs shrank in volume and grew more stagnant, oxygen levels fell, and temperature fluctuations became more severe. The cichlids struggled to find adequate spawning areas and oxygen-rich refugia to sustain their populations.

Pollution further compounded the crisis. Agricultural runoff, laden with pesticides and fertilizers, drained into the cichlids' already restricted waters. Sediments carried by irrigation channels began clogging the gravel substrates necessary for egg-laying. Meanwhile, introduced fish species—often stocked to diversify angling opportunities or to control pests—increased competition and predation pressures on the native Potosi Cichlid. Non-native tilapia, for instance, can outcompete local cichlids for food and spawning sites, while grass carp may alter the vegetation in spring-fed pools, reducing cover and shelter. In some springs, a combination of overfishing and invasive fish introductions spelled a rapid decline for the local cichlid assemblages.

By the 1980s and 1990s, the Potosi Cichlid was teetering on the brink. Sporadic conservation efforts, including petitions to protect spring habitats, were often overshadowed by local economic demands for water extraction. Authorities occasionally restricted well drilling or proposed minimal water flow regulations, but enforcement proved weak. Some scientists and conservationists tried to breed the cichlid in captivity, hoping to reintroduce it once habitats recovered. Yet captive breeding was complicated by the cichlid's specialized needs, and the rate at which water sources were degrading outpaced any efforts at restoration. Over time, a combination of water mismanagement, pollution, and competition from introduced species erased the Potosi Cichlid from its native range.

Although there have been occasional claims of sightings or reports of a handful of surviving individuals, these remain anecdotal. The consensus is that *Herichthys minckleyi* (in its original, distinct form) is now extinct in the wild, if not entirely lost. The extinction resonates beyond the fish itself, pointing to a pattern of aquifer depletion and wetland destruction in arid Mexico. Desert springs may appear marginal to some, but they are hotbeds of endemism, often containing species that evolved no analogues elsewhere on the planet. The

Potosi Cichlid's vanishing thus parallels the Tecopa Pupfish in California: both were desert-dwelling fishes that thrived in specialized, isolated water bodies, only to be driven to extinction when those fragile systems were altered by irrigation, tourism, or invasive species.

For local communities, losing the Potosi Cichlid means more than just a decline in biodiversity. The presence of these unique fish often attracted eco-tourists and biologists, creating potential for sustainable economic ventures. As the fish disappeared, so did a piece of cultural and natural heritage that could have served as a focal point for conservation and economic development. The tragedy underscores that short-term gains from water extraction or agricultural expansion can bring long-term losses that reverberate through ecosystems and livelihoods alike.

Main Causes: Pollution and Overfishing

From the steaming desert pools of the Tecopa Pupfish to the icy alpine depths once inhabited by the Gravenche, these extinct fish species illustrate the interplay of environmental degradation and unsustainable harvesting. While each case has its nuances—thermal habitat modification in Tecopa, damming and siltation in Lake Ontario's tributaries, and invasive species in Mexican desert springs—pollution and overfishing stand out as recurring, overarching threats.

Pollution

In nearly every scenario, the introduction of

contaminants—be they agricultural chemicals, industrial effluents, or urban wastewater—altered water chemistry and reduced the viability of fish populations. Eutrophication, brought on by an overload of nutrients, sparked cascading effects by triggering algal blooms, dropping oxygen levels, and suffocating fish eggs. Even the Tecopa Pupfish, adapted to extreme temperatures, could not withstand a shift in water quality triggered by tourism development. The Lake Geneva Whitefish and the Gravenche were similarly undone by nutrient-laden runoff that clouded their once-crystalline alpine waters. The Potosi Cichlid lost out to pesticides and fertilizers that leeched into desert springs, while the Lake Ontario Salmon faced a lethal mix of deforestation-driven sedimentation and industrial pollution. In each case, the essence of the fish's fundamental habitat—clean, well-oxygenated water—was compromised or eradicated altogether.

Overfishing

Alongside pollution, overfishing emerged as a powerful driver of decline. Fish populations can endure moderate harvest if given time and sufficient habitat for recovery. But when catches consistently outstrip reproduction, populations unravel. For species like the Lake Geneva Whitefish, advanced gear and rising market demand proved too great a burden. The Lake Ontario Salmon faced an onslaught of commercial and recreational fishing during critical spawning runs, effectively cutting off future generations. Gravenche and other European whitefish were similarly pursued beyond sustainable limits, undermining their capacity to rebound from pollution. Overfishing not only reduces adult population sizes but also removes the strongest breeders and disrupts social or migratory cues. Smaller, more fragmented populations are then more vulnerable to the slightest environmental changes or disease outbreaks.

Additional Stressors

Although pollution and overfishing are paramount, other factors amplified these impacts. Habitat fragmentation, for instance, was catastrophic for migratory species like the Lake Ontario Salmon, whose life cycle depended on free-flowing rivers to reach spawning grounds. Introduced species also played a role, as the Potosi Cichlid lost territory to non-native fish that competed for resources or preyed on juveniles. Government-led attempts to "help" fisheries through stocking often backfired, diluting native gene pools and further stressing wild populations. Tourism development led to channelization and chemical contamination in Tecopa Pupfish habitat, while industrialization produced sewage and industrial byproducts that fed into Lake Geneva, Lake Ontario, and other water bodies. These synergistic pressures ensured that once fish populations began declining, they had little hope of recovery unless swift and substantial mitigation measures were enacted.

Implications for Global Aquatic Biodiversity

The extinction events detailed here reflect a broader crisis affecting fish worldwide. As human populations expand and intensify their use of freshwater and marine resources, thousands of fish species face diminishing habitats, polluted waters, and relentless harvesting. Smaller endemic fish in desert springs, high-elevation lakes, or isolated river basins are especially prone to irreversible declines if their specialized conditions are disrupted. More broadly, commercial fisheries across the globe illustrate how quickly technologically advanced gear can empty seas and rivers that once seemed inexhaustible. Each local extinction also carries cultural, economic, and ecological ramifications—from the loss of traditional fishing livelihoods to the collapse of food webs and the disappearance of unique evolutionary histories.

Nevertheless, it is not too late to draw lessons from these tragedies. The protection of water quality through stringent pollution regulations, the creation of fish sanctuaries or "no-take" zones, and the implementation of sustainable fishing quotas all have proven track records of aiding fish population recovery. In some areas, dam removal or construction of effective fish passages has revived historically decimated runs of salmon, trout, and other migratory species. Habitat restoration work—replanting riparian zones, controlling invasive species, or rehabilitating degraded springs—can reestablish the ecological balance that fish need to thrive. Public awareness campaigns and economic incentives for conservation-minded tourism can shift community priorities away from short-term exploitation. Examples of success exist in places where local people, scientists, and governments cooperate to protect or rehabilitate fisheries.

Yet the stories of the Tecopa Pupfish, the Lake Geneva Whitefish, the Lake Ontario Salmon, the Gravenche, and the Potosi Cichlid warn us that once a species crosses the threshold into functional extinction, no amount of restoration can resurrect its original genetic uniqueness or cultural heritage. Extinction is permanent, even if new fish populations can be introduced later. Each loss diminishes the planet's biological wealth and the resilience of aquatic ecosystems under stress. With mounting challenges posed by climate change—rising water temperatures, altered precipitation patterns, and ocean acidification—fish species that are already near their tolerance limits could follow the path of those recounted in this chapter.

In sum, the extinction of these five fish species underscores how swiftly aquatic life can be eradicated when habitat quality declines, water is polluted, and fishing practices are unrestrained. Whether in a desert hot spring or an alpine lake, fish rely on the delicate interplay between water chemistry, temperature, oxygen levels, food webs, and breeding cycles. Pollution and overfishing strike at the very heart of these interactions. The window for preventing further losses may be closing, but by learning from these cautionary tales, policymakers, fishers, and communities can still act to protect many of the world's remaining fish species. The fate of global aquatic biodiversity will hinge on decisions made in the coming years—decisions about how we value and steward the waters that sustain not only fish, but the entire tapestry of life linked to rivers, lakes, springs, and seas.

CHAPTER 5: EXTINCT INSECTS

Insects represent by far the most diverse group of organisms on the planet, forming the foundation of countless ecosystems. They pollinate plants, recycle organic matter, provide food for other animals, and help control pests. Despite their abundance and pivotal ecological roles, insects are surprisingly vulnerable to sudden environmental changes. Many have highly specialized feeding, breeding, or nesting requirements that can be disrupted by habitat loss, pesticide use, climate fluctuations, and the introduction of invasive species. Over the last few decades, scientists have documented dramatic declines in insect populations worldwide, a phenomenon sometimes referred to as an "insect apocalypse." Although most insect extinctions often pass unnoticed by the wider public, given the small size and cryptic nature of these creatures, they are every bit as consequential for biodiversity as the loss of charismatic birds or mammals. This chapter spotlights five notable examples of extinct or presumed extinct insects: the Large Blue Butterfly (subspecies *Maculinea arion eutyphron*), the Mexican Bean Beetle (*Megacerus baeri*), the Xerces Blue Butterfly (*Glaucopsyche xerces*), the Greater Stag Beetle of Singapore (*Dorcus titanus sika*), and Franklin's Bumblebee (*Bombus franklini*). The chapter concludes with an analysis of two main drivers—pesticides and habitat loss—that have repeatedly emerged as the culprits behind these and many other insect extinctions.

6.1 Xerces Blue Butterfly (*Glaucopsyche xerces*)

Among extinct insects, the Xerces Blue Butterfly is one of the most iconic in North America. Endemic to the sand dunes and coastal scrub of the San Francisco Peninsula, *Glaucopsyche xerces* was known for its iridescent blue wings edged with a narrow row of white spots. This species had evolved in an area with mild, foggy summers and dunes dominated by native lupines, on which its caterpillars fed. Early European settlers described the dunes teeming with various butterflies and wildflowers, offering a natural spectacle at the edge of a growing urban center.

The downfall of the Xerces Blue can be traced largely to habitat destruction. San Francisco's rapid expansion in the 19th and early 20th centuries saw the leveling of dunes for housing, businesses, and military facilities. As roads, buildings, and ornamental landscaping replaced the sandy native vegetation, the lupines that served as larval host plants vanished. Meanwhile, the introduction of non-native grasses and shrubs further supplanted the specialized flora on which *G. xerces* depended. The butterfly's limited range—restricted to a small coastal strip—afforded little resilience once land development accelerated.

By the early 1940s, entomologists noted that *Glaucopsyche xerces* had become scarce. The final recorded sightings took place in the Presidio area of San Francisco, close to the shoreline where pockets of native dunes persisted. A few collectors captured specimens in hopes of establishing museum records, but there was no organized effort to protect or breed the butterfly. Soon afterward, *G. xerces* disappeared altogether, presumed extinct by the late 1940s or early 1950s.

In the years since, conservationists have elevated the Xerces Blue as a symbol of the fragility of specialized insects in urbanizing landscapes. The butterfly's name has also been adopted by the Xerces Society, a leading non-profit organization dedicated to invertebrate conservation worldwide. Attempts to "revive" something akin to *G. xerces* through captive breeding of closely related species have sparked debate. Some researchers argue that another coastal blue butterfly, the Silvery Blue (*Glaucopsyche lygdamus*), might fill a similar niche if dune habitats are restored. Yet the genetic lineage of the Xerces Blue itself is gone, underscoring that once a local population with unique evolutionary history is lost, it cannot be meaningfully replaced.

The lessons from the Xerces Blue resonate globally. Urban expansion remains a powerful driver of insect extinctions when natural areas, especially hotspots of endemism along coasts or river corridors, are viewed purely as prime real estate. Without deliberate efforts to preserve remnants of native vegetation, pollinators, herbivores, and predator insects vanish under roads and parking lots. And while some species adapt to city parks or ornamental gardens, many specialized insects do not. The plight of *G. xerces* thus stands as a stark reminder that uncontrolled development can erase centuries of evolutionary uniqueness in a matter of decades.

6.2 Franklin's Bumblebee (*Bombus franklini*)

Bumblebees are among the most recognizable pollinators in temperate ecosystems, revered for their fuzzy appearance and crucial role in fertilizing crops and wild plants. Franklin's Bumblebee (*Bombus franklini*) had one of the smallest geographic ranges of any North American bumblebee, confined primarily to a narrow zone of mountain meadows and agricultural valleys in southern Oregon and northern California. First described in the early 20th century, it was notable for its distinct coloration—a black abdomen with a few pale bands—setting it apart from more widespread relatives.

As with many bumblebees, *B. franklini* nested in underground or surface-level cavities, often in abandoned rodent burrows. The species relied on diverse floral resources from spring through fall, collecting nectar and pollen from wildflowers, shrubby plants, and, at times, cultivated crops like blueberries and raspberries. While its numbers were never large, *B. franklini* played a vital role in pollinating native plant communities. Local orchardists also benefited, since bumblebees can outperform honeybees in cooler, cloudier conditions prevalent at higher elevations.

Populations declined precipitously in the 1990s and early 2000s for reasons still debated. Researchers suspect multiple overlapping factors. The use of certain pesticides, especially neonicotinoids, rose in farmland adjacent to the bumblebee's range, potentially weakening colonies. Habitat fragmentation, spurred by housing developments and road expansions, disrupted nesting and foraging corridors. Climate change may have affected the timing of flowering or the frequency of late-spring frosts, reducing available blossoms. Another major concern was the spillover of pathogens from commercial bumblebee colonies used in greenhouse pollination. Diseases like Nosema or viruses can spread when domesticated bees escape or interact with wild populations, a well-documented issue in other regions.

Despite surveys by dedicated entomologists, sightings of *B. franklini* became exceedingly rare after the early

2000s. By the 2010s, many experts considered it either extinct or functionally extinct, as no breeding colonies had been identified for years. Conservation groups petitioned for federal protection, but the species' narrow range, small population, and the difficulty of immediate conservation interventions rendered efforts largely unsuccessful. Some bumblebees can be reared in captivity, but replicating the exact nest conditions and microclimates for a critically endangered species is challenging, especially if the cause of decline involves complex interactions of habitat loss, pesticides, and disease.

Today, *Bombus franklini* is commonly cited alongside the Xerces Blue Butterfly as a North American insect tragedy—an example of how quickly a specialized pollinator can vanish. The repercussions extend beyond the species itself: the plants that once depended on this bumblebee may see reduced pollination success, altering seed production and local plant community dynamics. The event also stirs broader reflection on the global pollinator crisis. While honeybees grab headlines due to their commercial importance, wild bees like *B. franklini* often suffer silently, overshadowed by more visible declines but equally crucial for ecosystem health.

6.3 Main Causes: Pesticides and Habitat Loss

The fates of the Large Blue Butterfly (*Maculinea arion eutyphron*), the Mexican Bean Beetle (*Megacerus *baeri*), the Xerces Blue Butterfly (*Glaucopsyche xerces*), the Greater Stag Beetle of Singapore (*Dorcus titanus sika*), and Franklin's Bumblebee (*Bombus franklini*) illustrate how insects, despite their evolutionary resilience, can be driven to extinction by two interwoven forces: pesticide use and habitat destruction.

Pesticides

From farmland to orchard to urban gardens, pesticides have become pervasive in modern ecosystems. In the case of the Mexican Bean Beetle, insecticidal drift from agricultural fields likely decimated both larvae and adults, even though the beetle rarely harmed domestic bean crops. Similarly, Franklin's Bumblebee lived in a region of increasing pesticide use, with systemic chemicals like neonicotinoids potentially affecting bumblebees' foraging behavior, immune systems, and breeding success. Pesticides can strike insects directly by poisoning them on contact or indirectly by contaminating their food plants, soil, and nesting materials. Even sublethal doses of systemic insecticides can impair orientation and reduce fertility, pushing populations below viable thresholds over time. Moreover, broad-spectrum pesticides often destroy the insects that pollinators or other beneficial species rely on for nutrition, unraveling complex ecological relationships. In the city context, pesticide usage in ornamental gardens and landscaping can further diminish insect refuges, leaving species nowhere to escape.

Habitat Loss

All five cases also underscore how crucial intact habitats are for insects. Whether it is the specialized grassland ecosystems required by the Large Blue Butterfly, the coastal dunes that sustained the Xerces Blue, or the temperate meadows favored by Franklin's Bumblebee, insects rely on microhabitats that supply specific host plants, nesting substrates, and climatic conditions. When these habitats are fragmented or replaced by urban sprawl, intensive agriculture, or industrial land uses, insect populations lose the spatial continuity needed for breeding and dispersal. For instance, the stag beetle in Singapore vanished as lowland forests were cleared and "tidied," removing the decaying wood that larvae needed. The Xerces Blue was squeezed out of existence by San Francisco's relentless coastal development. In farmland contexts, the push to eradicate "weedy" or "unproductive" vegetation spelled doom for insects like *Megacerus baeri*, which depended on patches of wild legumes.

Habitat fragmentation also compounds other threats. Populations restricted to small, isolated patches are more vulnerable to random events—droughts, storms, or disease outbreaks. Recolonization from neighboring patches becomes difficult or impossible if corridors are severed. Meanwhile, the microclimates that once moderated temperature and humidity can shift drastically in smaller fragments, subjecting insects to greater extremes. Even when modest efforts are made to create buffer zones or protected areas, these may prove insufficient if they are too small, poorly managed, or embedded in landscapes awash in pesticides.

Interactions and Synergies

Pesticides and habitat loss rarely act in isolation. In intensively managed agricultural landscapes, the elimination of native vegetation near fields removes both refuge habitats and beneficial insects that naturally control pests. Farmers, seeing fewer natural predators or pollinators, might be driven to apply more pesticides, which in turn further reduces insect diversity, creating a vicious cycle. On a broader scale, climate change can exacerbate these processes by shifting suitable habitat zones faster than insects can adapt or relocate. In the case of high-altitude species such as *Maculinea arion eutyphron*, warming temperatures can push them into smaller pockets of cooler habitat, heightening the impact of land-use changes. If pesticide use also extends into these last refuges, extinction probabilities soar.

Conservation Responses

Preventing future insect extinctions requires a multi-pronged approach. Reducing pesticide reliance—through integrated pest management, organic farming, and targeted pesticide application rather than blanket spraying—can lessen the toxic load. Conservation planners emphasize setting aside protected areas that remain pesticide-free, enabling insects to breed and forage in relatively undisturbed habitats. Restoring native flora is vital: specialized pollinators, brood parasites, and herbivores each need their respective host plants or nesting substrates. A mosaic of land-use types, including buffer strips of wildflowers along farm edges, "untidy" corners with fallen logs, and well-managed nature reserves, can help re-establish ecological corridors and networks. Where possible, local and national regulations limiting pesticide usage—especially the most lethal classes—have shown promise for stemming insect declines.

Education and citizen science also hold potential. Many insects go extinct before they're even identified or studied, so training amateur naturalists to monitor local insects can lead to earlier detection of declines and a stronger impetus for conservation. Public appreciation for pollinators, once mostly directed at honeybees, has broadened to include bumblebees, solitary bees, and other crucial groups, spurring projects like backyard pollinator gardens and community-based habitat restoration.

Yet the stories in this chapter remind us that for some insects, these actions have come too late. The Xerces Blue in San Francisco's dunes and Franklin's Bumblebee in the mountain meadows of the Pacific Northwest stand as stark cautionary examples: a specialized pollinator or herbivore can collapse rapidly once its environment is sliced apart and drenched in chemicals. The same fate befell the specialized grassland butterfly in Europe or the wild-bean-dependent beetle in Mexico. Each extinction narrows life's tapestry, removing evolutionary marvels that will not re-emerge. And while certain insect declines might initially go unnoticed, their ripple effects in ecosystems—such as loss of pollination or shifts in plant community structure—will eventually concern human populations that depend on stable agricultural systems and functional natural habitats.

Final Reflections

Insects carry out essential services: pollination, pest control, soil aeration, nutrient cycling, and the creation of habitats for myriad other organisms in the food web. Their extinctions—quiet, often unseen—reflect broader patterns of ecological upheaval driven by anthropogenic forces. By tracing the demise of the Large Blue subspecies, the Mexican Bean Beetle, the Xerces Blue, the Greater Stag Beetle of Singapore, and Franklin's Bumblebee, we see the same grim themes repeat: expanding farmland or urban sprawl replaces the mosaic of natural vegetation; pesticides disrupt or poison the fragile life cycles of specialized insects; and insufficient conservation attention permits species to slip away largely undocumented. If there is a silver lining, it lies in the growing awareness of insect declines globally. Efforts to reduce pesticide dependence and protect remaining habitats have taken root in many regions, fueling optimism that other insects teetering on the brink might yet be saved. The disappearance of these five species underscores what is at stake—a world with fewer colors, fewer pollinators, and fewer wonders, should we fail to safeguard the small creatures that underpin so much of our planet's vitality.

CHAPTER 6: IMPACT ON ECOSYSTEMS

Extinction never occurs in a vacuum. While each vanished species signifies the loss of a distinct evolutionary lineage, there is also a ripple effect that echoes through the ecological communities where that species once played a role. Whether the missing organism was a top predator, a pollinator, a seed disperser, a nitrogen fixer, or any of countless ecological functions, its absence alters the dynamics of food webs and nutrient cycles. This chapter delves into how the extinction of species—be they mammals, birds, reptiles, amphibians, fish, or insects—reshapes ecosystems. It focuses on three interconnected phenomena: the domino effect that triggers imbalances in natural communities, the consequent loss of biodiversity that reduces ecosystem resilience, and the ensuing disruptions in food chains. Throughout these sections, we will see why every extinction reverberates beyond the disappearance of a single organism, carrying implications for entire habitats and, ultimately, for human well-being as well.

7.1 The Domino Effect on Natural Balance

Ecological systems are intrinsically interconnected, more akin to a web of intricate feedback loops than a simple chain of direct cause-and-effect. When one species vanishes, the repercussions can fan out in complex and sometimes unpredictable ways—hence the "domino effect." Although not all extinctions cause equally visible cascades, many species play roles crucial enough that their loss or decline triggers noticeable, even dramatic, reverberations.

7.1.1 Keystone Species and Trophic Cascades

A core concept illustrating the domino effect is that of the **keystone species**, organisms that have a disproportionately large impact on their environment relative to their abundance. Classic examples include apex predators—like wolves and sharks—and ecosystem engineers—like beavers and large herbivores. Removing a keystone species can lead to **trophic cascades**, where changes in the predator or engineer level affect the abundance of species at lower or higher trophic levels, reconfiguring the entire food web.

A pertinent historical example is the extirpation of wolves from many regions of North America. With the disappearance of this apex predator, populations of elk and deer soared, decimating young trees and forest undergrowth, and changing riverbank vegetation so dramatically that erosion patterns shifted. Reintroducing wolves to Yellowstone National Park famously reversed some of these changes, illustrating how a single predator's absence or presence can exert sweeping control over entire ecosystems. Similarly, in marine environments, the loss of large sharks can allow mesopredators—like rays and smaller shark species—to proliferate unchecked, in turn reducing populations of bivalves or crustaceans upon which they feed, and so forth.

7.1.2 Ecosystem Engineers and Habitat Reshaping

Keystone species may also be **ecosystem engineers** that modify their environments in ways that benefit myriad other organisms. Elephants, for instance, push over trees to create open grasslands; beavers build dams that transform flowing streams into wetlands. When these engineers disappear, the landscapes they once shaped begin to revert to different successional states. For example, the extinction of giant tortoises on many Indian Ocean islands meant the end of critical seed dispersal for certain plant species, which subsequently declined or failed to reproduce effectively. Over time, vegetation density and species composition shift, impacting insects, small vertebrates, and eventually the entire ecological community.

7.1.3 Mutualistic Networks and Plant-Pollinator Interactions

The domino effect also appears in mutualistic networks such as pollination. Pollinator declines—exacerbated by the disappearance of certain insects or birds—can reduce the reproductive success of the plants they service. When specific pollinators vanish, specialized flowers that rely heavily on them may produce fewer seeds, gradually diminishing their population. In turn, other animals that depend on these plants for food or nesting materials face shortages. In some tropical forests, the loss of large-bodied fruit-eating birds has caused an uptick in trees with smaller fruits (dispersed by smaller birds or bats) and a decline in large-fruited species. Such shifts can tip the balance of entire forest compositions, possibly influencing carbon storage, soil fertility, and even the microclimates within the forest.

7.1.4 Invisible Ties in Microbial and Soil Communities

While charismatic fauna garner the most attention, the disappearance of species also shapes microbial and soil communities in ways that can be subtle but profound. Soil organisms such as mycorrhizal fungi or nitrogen-fixing bacteria often rely on the organic matter or root interactions facilitated by diverse plant and animal life. When a particular seed-dispersing bird or mammal disappears, the distribution of seeds changes, altering which plants dominate the understory or canopy. This vegetation shift in turn modifies the soil chemistry, hydrology, and microclimate that soil microbes experience, leading to potential knock-on effects for decomposition rates and nutrient cycling. These microbial shifts, though less visible, can cascade back up the food web by influencing how well plants grow, which in turn affects herbivores and higher-level predators.

7.2 Loss of Biodiversity

While many people think of biodiversity as the sheer number of different species in a given area, ecologists define it more broadly to encompass the genetic diversity within those species, the variety of functional traits in a community, and the range of habitats that comprise an ecosystem. Each extinction reduces some aspect of this diversity—sometimes in ways that are immediately apparent, and sometimes in ways that only become evident over longer temporal scales.

7.2.1 Reduced Ecosystem Resilience

One of the most significant consequences of losing biodiversity is a decline in **ecosystem resilience**—the ability of an ecosystem to withstand and recover from disturbances such as fires, floods, storms, or disease outbreaks. A highly diverse community often can buffer against shocks because multiple species may fill similar ecological niches or provide overlapping functions. If one pollinator species declines due to a disease, a different pollinator might temporarily step in. This "insurance" effect, however, weakens as species vanish. When a disturbance arrives in a less diverse system, the remaining species may lack the functional redundancy to compensate, potentially leading to a total system collapse or a regime shift into a less desirable state. For example, a coral reef heavily dominated by a single coral species might swiftly crumble if a bleaching event kills that species, whereas a reef with multiple resilient coral types might partially recover.

7.2.2 Genetic Erosion and Adaptation

Biodiversity also encompasses **genetic variation** within species. A population with high genetic diversity typically displays greater adaptability to environmental changes like temperature fluctuations, diseases, and shifts in resource availability. If that species experiences a severe population bottleneck due to habitat loss or overexploitation, it loses much of that genetic repertoire, becoming less capable of adapting to future pressures. Such genetic erosion can spiral into an "extinction vortex," where small populations become more inbred and thus more vulnerable to random events.

This trend plays out vividly among large mammals—such as rhinoceroses, tigers, or pandas—that have been driven down to just a few dozen or hundred individuals. Even if these populations temporarily rebound, the genetic diversity they regain is limited. This constraints their potential resilience to new pathogens or climate change. Conversely, once-extinct species cannot evolve at all—any unique genetic traits they possessed are lost forever, wiping out millions of years of evolutionary innovation in the blink of a geological eye.

7.2.3 Cultural and Aesthetic Losses

Though often overlooked in scientific discourse, the loss of biodiversity also robs humanity of cultural, aesthetic, and even spiritual values. Many communities identify deeply with the local animals and plants that mark seasonal changes or feature in folklore. Extinctions of birds, mammals, or insects can erode traditional knowledge built up over centuries, severing cultural ties. Tourism can also suffer, particularly in areas where iconic species (like tigers, gorillas, or whales) draw visitors. Furthermore, the dissolution of culturally significant relationships, such as those between indigenous peoples and the wild game or medicinal plants they steward, can lead to broader socio-economic challenges and a weakening of conservation ethics at the local level.

7.2.4 Domino Effect on Human Well-Being

While the initial sections focused on ecological networks, biodiversity loss eventually loops back to humans in myriad ways. The disappearance of pollinators, for example, can threaten agricultural production, particularly for fruits, nuts, and vegetables that rely on insect or bird pollination. As pollinator diversity dwindles, farmers may resort to expensive artificial pollination or see yields plummet, affecting food security and the economy. Similarly, overfishing can collapse fisheries that once sustained coastal communities, driving unemployment and potentially catalyzing conflicts over dwindling resources. In the end, biodiversity functions as a natural capital—its depletion can destabilize livelihoods and reduce resilience to environmental shocks such as droughts, floods, or disease epidemics.

7.3 Impact on Food Chains

Food chains describe linear sequences of who eats whom, whereas **food webs** offer a more realistic mosaic of interconnected predation, competition, and mutualism. The extinction of any single link can unravel multiple chains, producing a cascade of ecological consequences. The severity and nature of these impacts depend on whether the missing species occupied a position as a primary producer, a consumer, or a top predator.

7.3.1 Bottom-Up Effects

When a plant species goes extinct or drastically declines, a **bottom-up** effect can ripple through herbivores and, subsequently, the predators that rely on those herbivores. Primary producers form the base of most food webs, capturing energy from the sun through photosynthesis or from chemical processes in some specialized ecosystems (e.g., hydrothermal vents). If a particularly abundant or nutritionally essential plant disappears, the herbivorous insects or ungulates that fed on it must adapt by switching to different plants—if such alternatives exist. Some herbivores, however, may be too specialized to switch, leading them to suffer population declines. Predators that rely heavily on those herbivores either shift prey targets or face their own decline. Over time, entire communities can shift their composition and structure, with opportunistic or invasive species stepping in to fill vacated niches.

7.3.2 Top-Down Effects

Top-down control becomes evident when apex predators or keystone predators vanish. In terrestrial systems, the removal of wolves, big cats, or large raptors often results in prey overabundance. The subsequent overgrazing or overbrowsing can degrade vegetation, reduce habitat for smaller animals, and even affect soil stability and water quality. The disappearance of large fish from a coral reef or coastal ecosystem can prompt mesopredator release—shifting the balance among medium-sized fishes and invertebrates like sea urchins, which may then overgraze coral algae, hindering coral recruitment.

Marine systems in particular showcase how removing top predators can fundamentally alter ecosystems. Cod collapses in the North Atlantic led to a proliferation of smaller forage fish, which in turn suppressed the recovery of cod by feasting on cod larvae and juveniles. Attempts to "fish down" the food web can produce short-term spikes in catches, but often lead to long-term instability, where large, commercially valuable fish are replaced by jellyfish or smaller, less desired species.

7.3.3 Mid-Level and Specialist Species

Not all species fit neatly into the concept of apex predator or foundational primary producer. Many occupy middle tiers or serve specialized roles, such as seed dispersers, pollinators, or nest builders for other creatures. Losing these mid-level or specialist species can break important links in food webs. For instance, certain frugivorous birds or bats may be responsible for spreading the seeds of dozens of tree species. If those birds or bats decline, tree recruitment and forest regeneration suffer. Likewise, the extinction of key pollinators in mountain meadows or desert ecosystems can drastically reduce the reproduction of certain flowering plants, subsequently impacting herbivores that rely on those plants. Each shift may not be as dramatic as a top-down or bottom-up collapse, but cumulatively, such changes reshape the tapestry of life, often leading to simpler, less productive ecosystems.

7.3.4 Emergence of Pest or Invasive Species

Some extinctions can inadvertently open the door for pest or invasive species to thrive. When predatory insects vanish, invasive herbivorous insects might proliferate, damaging crops or native vegetation. Alternatively, if a native herbivore disappears, an invasive herbivore with no local predators or competitors might fill the vacant niche, often in more destructive ways. On small islands where reptiles like giant tortoises or flightless birds once kept certain plants and pests in check, their loss can trigger explosive growth of alien rodents, cats, or invasive plants, perpetuating a vicious cycle that leads to further extinctions. The net effect is a more homogenized ecosystem, dominated by a few invasive generalists rather than the rich array of specialists that evolved over millennia.

A Holistic Understanding of Extinction's Impact

Synthesizing these elements reveals why scientists and conservationists stress the interconnectedness of life. Extinctions, whether of large mammals or minute insects, have cascading impacts that reshuffle species interactions, degrade ecosystem functions, and reduce the overall resilience of natural communities. From the perspective of ecological dynamics, each species is part of a larger puzzle—removing one piece may weaken the entire structure, even if the immediate consequences aren't always visible to casual observation.

7.3.5 "Ghost Ecosystems" and Functional Extinctions

An important concept relevant here is that of "ghost ecosystems," where the structure of an environment superficially remains, but the key processes that once kept it functioning are gone or severely diminished. For instance, a forest might still stand, but if pollinators, seed dispersers, or decomposers have been lost or reduced, the forest is no longer able to regenerate effectively. It becomes a ghost of its former self, inevitably aging and losing species until it transitions into a different, often impoverished state. This phenomenon is linked to "functional extinction," where a species may still have a few surviving individuals but no longer plays its ecological role effectively.

7.3.6 Resilience, Adaptation, and Potential Reversibility

Ecosystems do exhibit levels of resilience and can sometimes adapt to extinctions over long timescales. New interactions emerge, or surviving species might evolve or fill empty niches. Nevertheless, the relatively rapid pace of anthropogenic extinctions poses a serious challenge. Natural adaptive processes operate over thousands to millions of years, whereas human-driven habitat changes, overharvesting, and climate shifts unfold within decades. This mismatch in timescales results in diminished adaptability and an increasing likelihood of collapse rather than a neat readjustment.

Occasionally, conservation interventions—such as reintroducing a lost predator or herbivore—offer a glimpse of partial reversibility. Yellowstone National Park, mentioned earlier, witnessed a rebound in riparian vegetation, beaver colonies, and overall biodiversity after wolf reintroduction. Wetland restorations that reintroduce beavers or waterfowl can revive certain ecosystem functions. Such cases suggest that, under the right conditions, targeted actions can repair at least some of the damage caused by extinctions. However, true reversibility is rare when a species is fully extinct, as no amount of ecosystem repair can resurrect a lineage that no longer exists. Reintroduction is possible only if a genetically and ecologically compatible surrogate or closely related species remains.

Conclusion: Learning from Cascades

Every extinction sends waves through the living world. Whether it's a large-bodied mammal that once roamed across continents or an unassuming insect that pollinated a select group of flowers, its disappearance alters the webs of interactions sustaining life. The loss of keystone species can spark drastic changes in ecosystem structure, while the loss of pollinators or seed dispersers erodes biodiversity and forest regeneration. Plants, microbes, and soils shift in composition, reconfiguring how energy and nutrients flow through the system. Over time, repeated extinctions reduce the ecological richness that once allowed communities to resist and adapt to environmental changes.

Understanding the multifaceted effects of extinction reminds us why conservation is not simply about preserving individual species for their inherent or aesthetic value. It is also about safeguarding the underlying processes—pollination, seed dispersal, water purification, predator-prey regulation, and more—that keep ecosystems functioning. When these processes collapse, the costs are borne not just by wildlife but by human societies dependent on stable fisheries, fertile soils, and reliable pollinators for crops.

From a policy standpoint, recognizing how extinctions cascade through ecosystems underscores the necessity of **ecosystem-based management**. This approach focuses on preserving entire communities and the complex links within them, rather than treating species in isolation. Initiatives might include maintaining habitat corridors that support migratory species, regulating hunting or fishing to sustainable levels, controlling invasive species that disrupt native webs, and prioritizing the restoration of ecological functions. Protecting the top predators, pollinators, and ecosystem engineers can prove especially effective, ensuring that the keystone interactions remain intact.

Nevertheless, the challenge remains formidable. Climate change accelerates habitat shifts, introducing new uncertainties. Economic pressures on land and water use intensify resource extraction, while globalization facilitates the spread of invasives. Addressing these threats requires international cooperation, robust environmental legislation, and incentives that value biodiversity's long-term benefits. It also entails that local communities, scientists, and policymakers converge on solutions grounded in both scientific knowledge and traditional ecological insights.

In sum, the impact of extinction on ecosystems extends far beyond the absence of a particular species. It can set off a chain of transformations that weaken biodiversity, disrupt food webs, and undermine the very services upon which life—including human life—depends. By appreciating these domino effects and acting promptly, societies can help prevent cascading failures in the natural world. Conservation thus emerges not merely as an ethical duty to preserve life's variety, but also as a pragmatic strategy to maintain the ecological resilience needed to face the uncertainties of a rapidly changing planet.

CHAPTER 7: WHAT WE CAN DO

The preceding chapters have illustrated the breadth and depth of the modern extinction crisis—from the disappearance of iconic mammals and birds to the vanishing of less-celebrated reptiles, amphibians, fish, and insects. We have seen that each extinction reverberates through ecosystems, eroding biodiversity and compromising the resilience of the natural world upon which human societies also rely. Yet if the story of extinction is sobering, it is not entirely one of despair. Around the globe, individuals, communities, organizations, and governments are working creatively and persistently to halt, and sometimes reverse, the losses we face. This chapter explores strategies, solutions, and everyday actions that can shift the trajectory away from biodiversity collapse and toward a more harmonious coexistence with other species.

A crucial recognition underpins these efforts: the fate of wildlife and ecosystems is not determined solely by environmental factors but by human decisions—political, economic, cultural, and ethical. The power to cause extinctions is, after all, the obverse of the power to prevent them. Whether one is a policy-maker drafting legislation, a conservation biologist restoring a wetland, a teacher integrating conservation themes into lesson plans, or simply a citizen making daily choices about what to eat or how to travel, each person holds a piece of the puzzle. Our collective challenge lies in coordinating these pieces, ensuring that conservation initiatives are robust, inclusive, equitable, and grounded in science. The sections that follow outline four broad areas of action: formal conservation measures to protect endangered species, habitat restoration to rebuild the stage on which wildlife depends, education and awareness to foster societal support, and daily personal steps that can aggregate into significant positive change.

8.1 Conservation and Protection of Endangered Species

8.1.1 Legal Frameworks and International Agreements

One of the most direct mechanisms for preventing extinction is legal protection at national and international levels. Treaties and conventions—such as the Convention on International Trade in Endangered Species of Wild Fauna and Flora (CITES), the Convention on Biological Diversity (CBD), and the Ramsar Convention on Wetlands—provide frameworks for countries to collaborate on conservation goals. At a national scale, legislation such as the U.S. Endangered Species Act or the European Union's Habitats Directive can enforce regulations that protect critical habitats, regulate hunting and trade, and mandate recovery plans for listed species.

These laws can be powerful. The Endangered Species Act, for example, has brought many iconic species in the United States—such as the Bald Eagle or the Gray

Wolf—back from the brink. Yet legislation is only as effective as its enforcement. Corruption, lack of funding, or insufficient political will can weaken protective measures. In some countries, wildlife laws exist on paper but rarely translate into real-life action. Moreover, not all threatened species are charismatic or well-known enough to rally political support. Many invertebrates, plants, and small fish remain unprotected because they do not capture public imagination or because baseline data on their status is lacking.

International agreements likewise face hurdles. CITES, intended to regulate or ban trade in endangered species and their products, depends on each member country's ability and desire to enforce strict border controls and penalize traffickers. When demand for ivory, rhino horn, or tiger bones remains high on the black market, poaching syndicates exploit legal loopholes or corruption. Hence, legal tools must be coupled with strong institutional capacity, transparent governance, and community engagement to curb illegal wildlife trade effectively.

8.1.2 In Situ Conservation: Reserves, Parks, and Community-Led Initiatives

In situ conservation—protecting species in their natural habitats—remains the cornerstone of biodiversity preservation. National parks, wildlife reserves, and marine protected areas (MPAs) can safeguard large tracts of land or ocean from destructive activities. Such protected zones ideally preserve not only single species but entire ecological communities. When well-managed, they allow natural processes like predation, pollination, and nutrient cycling to continue.

Yet the mere declaration of a protected area does not guarantee its success. Adequate funding and staff, robust monitoring, and respectful collaboration with local communities are essential. If local residents view a reserve as an imposition that restricts their livelihoods without offering benefits, conflicts inevitably arise. By contrast, community-led conservation initiatives often yield greater success and long-term stability. Indigenous groups, for instance, have stewarded biodiversity-rich territories for centuries. Recognizing indigenous rights and knowledge can strengthen conservation outcomes, as these communities have vested interests in sustainably managing their ancestral lands.

Co-management arrangements—where governments, NGOs, and local communities share responsibility—have proven effective in many contexts. Such models ensure that decisions about resource use, patrolling, and tourism are made collaboratively. They also open up avenues for local people to gain economic benefits—through eco-tourism or sustainable harvest of certain resources—while maintaining ecological integrity. A healthy synergy between traditional ecological knowledge and scientific research can enrich management strategies, broadening the adaptive capacity of conservation plans.

8.1.3 Ex Situ Conservation: Zoos, Seed Banks, and Gene Repositories

When a species is perilously close to extinction, ex situ measures might be the only option. Zoos, botanical gardens, aquariums, and seed banks act as repositories for genetic material. Captive breeding programs aim to stabilize or increase population numbers while scientists address the threats in the wild. The Arabian Oryx, California Condor, and Mauritius Kestrel are well-known species rescued by such programs. For plants, seed banks like the Millennium Seed Bank store seeds from tens of thousands of species, serving as a form of insurance against habitat destruction or climate catastrophes.

Ex situ conservation is not without complications. Captive environments can cause animals to lose behaviors needed for survival in the wild—such as migratory patterns, foraging tactics, or anti-predator responses—making reintroduction risky. Genetic diversity can be lost if the captive population is too small or poorly managed. Moreover, ex situ initiatives are costly and require long-term commitment, raising ethical and economic questions about resource allocation. Should extensive funding go into saving a single species ex situ when multiple habitats and species could benefit from broader ecosystem-level investments? There is no universal answer, but a balanced approach often proves most rational: ex situ programs function best when integrated with habitat conservation, threat mitigation, and community involvement.

8.1.4 Combating Poaching and Illegal Wildlife Trade

For many large mammals—elephants, rhinos, big cats—and certain reptiles—like turtles and tortoises—poaching is a major driver of decline. Even for lesser-known species like pangolins, incessant demand in illegal markets exacts a heavy toll. Addressing poaching requires more than field patrols. It entails dismantling international trafficking networks, employing advanced forensic techniques to track wildlife products, and applying technology like drones or GPS collars to aid enforcement. Equally important are community-based approaches: when local people receive economic alternatives, education, and a sense of ownership over wildlife resources, they often become the most effective guardians against poachers.

Demand reduction campaigns in consumer countries are crucial. Many wildlife products—rhino horn, tiger bones, pangolin scales—are used in traditional medicines, perceived as status symbols, or collected for exotic pet markets. Public awareness campaigns, social marketing, and strict legal penalties can shift cultural norms and discourage consumption. Progress is noticeable in certain regions—some surveys show reduced demand for ivory in China and Vietnam following high-profile awareness efforts. Yet persistent myths, entrenched customs, and profiteering networks mean that combating wildlife crime remains an ongoing battle requiring global cooperation.

8.2 Restoring Habitats

8.2.1 The Importance of Habitat Quality

Even the most robust protection programs cannot succeed if the habitats species depend upon are degraded or gone. Habitat restoration, therefore, is a pivotal complement to conservation measures focused on individual species. Healthy habitats not only support wildlife but also sustain human communities—filtering water, stabilizing soils, controlling floods, and sequestering carbon. In essence, restoring habitats is an investment in the ecosystems that support all life, including ours.

Habitat degradation can take many forms: deforestation, wetland drainage, coral reef bleaching, desertification, river damming, and more. Restoration aims to reverse or mitigate these processes, returning landscapes and seascapes to a state of ecological functionality. This is often a complex task. It may mean planting native trees, removing invasive species, reintroducing keystone animals, cleaning up pollutants, or re-wetting drained peatlands. The specific actions vary depending on the ecosystem's type and the severity of degradation.

8.2.2 Reforestation and Afforestation

Forests are among the planet's most biodiverse terrestrial habitats, and they play a vital role in regulating climate through carbon storage. Restoring deforested or heavily logged areas can quickly boost biodiversity if undertaken with care. Reforestation sometimes involves planting seedlings of native species, ensuring that a mix of tree types creates structural and species diversity. In other scenarios, simply protecting a site from further disturbance allows natural regeneration to occur, aided by seed banks in the soil or dispersal by animals. Each approach has pros and cons. Tree planting can guide the composition of the emerging forest, but it may be labor-intensive and prone to failures if seedlings are mismatched to local conditions. Natural regeneration harnesses nature's own processes but can be slow or uneven.

In some regions, **afforestation**—the planting of forests where none existed previously—is a controversial strategy. It can be beneficial if degraded grasslands or marginal farmland are planted with native forest species, boosting carbon sequestration and creating wildlife corridors. But if exotic monocultures or non-native species are introduced merely to maximize timber yields or carbon credits, then biodiversity benefits may be low. Indeed, planting the wrong species in grassland or savanna ecosystems can degrade unique habitats and displace specialized wildlife. A thorough understanding of the historical ecology and climate of a region is thus essential when planning large-scale reforestation or afforestation.

8.2.3 Wetland and Coral Reef Restoration

Wetlands (marshes, swamps, bogs, mangroves) are incredibly productive ecosystems, hosting a wide array of fish, amphibians, birds, insects, and plants. They also serve as natural water filters, removing pollutants and buffering floods by absorbing excess water. Restoring degraded wetlands typically involves removing drainage infrastructure, controlling invasive plants, and sometimes reintroducing beavers or other ecosystem engineers to reshape hydrology. Once water flow is restored, native wetland vegetation can recolonize and support a revival of animal life, from waterfowl to amphibians and invertebrates.

Similarly, coral reefs face grave threats from warming seas, ocean acidification, overfishing, and pollution. Restoration efforts may range from establishing marine protected areas that curb destructive fishing practices to more direct interventions, like transplanting fragments of resilient coral species onto artificial structures. Some scientists experiment with breeding coral strains better adapted to higher temperatures, hoping to create "super corals" that can survive warming oceans. Though promising, such interventions can be expensive and labor-intensive, and they merely buy time if global carbon emissions continue unabated. Ultimately, reef restoration must be coupled with broader climate action and better coastal management to stand a real chance of success.

8.2.4 Rewilding and Species Reintroductions

Rewilding is an ambitious form of habitat restoration that emphasizes minimal human intervention, allowing natural processes to drive ecological recovery. A hallmark of rewilding is the reintroduction of keystone or "apex" species that once roamed an area but were hunted or driven out. Examples include bringing wolves back to parts of Europe and North America, or reintroducing European bison to forests where they had disappeared centuries ago. By restoring the interactions between predators and prey, rewilding can catalyze broader habitat transformations—reducing overgrazing, fostering tree and shrub regeneration, and reshuffling competitive relationships among species.

Of course, rewilding faces social, economic, and logistical challenges. Local residents may fear large predators or resent restrictions on land use. Livestock owners worry about predation. Ecological uncertainties linger: how well can a predator adapt to a landscape now fragmented by roads and farmland? Some critics argue that rewilding overlooks the role of humans, especially indigenous peoples, who have long shaped landscapes. Others see it as a chance to expand the scope of traditional conservation, moving beyond small, fenced reserves to restore dynamic processes across entire regions. When done collaboratively, rewilding can revitalize ecosystems while also creating opportunities for eco-tourism, scientific study, and cultural revival.

8.2.5 Balancing Restoration with Local Needs

All habitat restoration projects interact with human communities living in or around the targeted areas. If local people are excluded from decision-making, or if restoration disrupts livelihoods without offering viable alternatives, tensions arise. Successful restoration therefore integrates socio-economic components. Projects might employ local inhabitants for tree planting, ecotourism guiding, or invasive species removal. They may also create educational programs to connect communities with the restored environment, fostering a sense of ownership and pride. When restoration aligns with local aspirations—such as securing clean water or providing new income sources—it tends to be more enduring.

Moreover, restoration must sometimes accommodate forms of sustainable resource use. Strict no-take zones or rewilding areas may be critical, but buffer zones permitting regulated harvesting of timber, fish, or non-timber forest products can reconcile conservation goals with economic realities. Ultimately, bridging ecological targets with human well-being lies at the heart of successful habitat restoration. It acknowledges that healthy environments and healthy communities go hand in hand.

8.3 Education and Awareness

8.3.1 The Power of Knowledge

Public awareness is a driving force behind conservation success. When people understand why species are endangered and how biodiversity underpins ecosystem services, they are likelier to support protective measures, donate to conservation causes, or alter behaviors that harm wildlife. Education spans all ages and social contexts, from primary schools to universities, from local communities to global media campaigns. The more widely the message is disseminated—that extinction has real and tangible consequences—the greater the momentum for political and economic change.

Education is especially pivotal in regions where human-wildlife conflicts occur. Villagers living near elephant habitats, for instance, may resent crop-raiding herds if they see no direct benefit to conservation. Effective education can demonstrate how ecotourism or compensation schemes might offset losses and emphasize the irreplaceable ecological services provided by elephants, such as seed dispersal. Similarly, fishermen might oppose marine protected areas until they learn how fish nursery grounds enable sustainable fisheries in the long run. Information, delivered respectfully and through trusted channels, can shift attitudes from antagonism to cautious cooperation.

8.3.2 Formal Education Systems

In many countries, curricula now incorporate environmental studies or ecology modules, teaching students about endangered species and conservation challenges. Some schools partner with local conservation groups to create field-based learning experiences, such as visiting nature reserves or assisting in citizen science projects. These hands-on encounters can kindle a lifelong appreciation for nature. However, not all educational systems prioritize environmental literacy. Budget constraints, competing subjects, and the lack of teacher training can limit effectiveness. Advocates for biodiversity awareness often push for curriculum reforms and teacher development programs, ensuring that future generations grow up with a nuanced understanding of ecological interconnectedness.

Outside the classroom, youth organizations—like Scouts, 4-H clubs, or nature clubs—offer further opportunities for experiential learning. Young people might participate in bird counts, beach clean-ups, or adopt-a-stream projects. Such engagements do more than build environmental knowledge; they foster leadership, teamwork, and a sense of responsibility for one's local environment. Over time, these experiences can translate into broader advocacy for sustainability.

8.3.3 Community-Based Initiatives and Citizen Science

Community-based education operates on the principle that local knowledge and participation are critical. By involving villagers, farmers, or fishers in biodiversity monitoring—for instance, collecting data on nesting birds or reporting sightings of rare species—citizen science programs can gather valuable data over wide areas. Simultaneously, these projects raise ecological awareness, empowering participants to become frontline conservationists. This approach has worked particularly well for migratory species such as monarch butterflies, sea turtles, and whales, whose vast ranges necessitate broad public engagement.

In community-led projects, elders can share traditional ecological knowledge—insights gleaned over generations about seasonal patterns, wildlife behavior, and sustainable resource use. Such knowledge, when combined with modern scientific methods, offers a richer perspective on ecosystem dynamics. Education then becomes a two-way dialogue, not a top-down lecture from external experts. Villagers who see the tangible benefits of species protection—perhaps through improved crop yields when natural pollinators rebound—are more inclined to champion conservation, passing their enthusiasm onto neighbors and future generations.

8.3.4 Media, Campaigns, and Storytelling

Beyond formal and community education, the media plays a formidable role in shaping public attitudes toward extinction. Documentaries like those produced by the BBC or National Geographic can transport viewers into the heart of remote ecosystems, forging emotional connections to distant wildlife. Environmental journalists highlight success stories—such as the comeback of certain whales or raptors—proving that conservation works. Simultaneously, investigative reporting exposes poaching rings, habitat destruction, or corruption, spurring international pressure.

Social media amplifies these effects. Platforms like Instagram, TikTok, or Twitter allow conservationists and organizations to reach global audiences rapidly. Eye-catching images of endangered species or short videos of rehabilitated animals can spark viral interest. Crowdfunding campaigns raise money for local reserves, while digital petitions press government agencies to protect habitats. Yet, social media can also spread misinformation or oversimplified narratives, necessitating critical thinking and fact-checking. Good storytelling remains key: if people feel empathy and hope, they are more likely to engage and support conservation initiatives.

8.3.5 Fostering a Culture of Conservation

Ultimately, education and awareness aim to cultivate a **culture of conservation**, where safeguarding biodiversity is woven into collective values and daily decisions. Such a culture does not emerge overnight. It grows through consistent messaging in schools, community programs, media, and policy discourse. Leaders, influencers, and role models—celebrities, religious figures, sports icons—can shift social norms by publicly endorsing wildlife protection or sustainable living. Over time, new traditions may form: local festivals celebrating migratory birds, civic pride tied to reforested landscapes, or personal milestones marked by adopting an endangered species. Where conservation is embraced not just as a scientific or policy imperative but as a shared moral and cultural principle, the path to preventing further extinctions becomes far more stable.

8.4 Daily Actions for Sustainability

8.4.1 The Individual's Sphere of Influence

While large-scale systemic shifts—strong environmental legislation, corporate responsibility, international agreements—are indispensable, individual behavior also matters. Each choice we make—from the food we eat to the products we purchase—carries an environmental footprint. These small-scale actions, replicated across millions or billions of consumers, can drive market trends, reshape corporate priorities, and reduce the strain on habitats worldwide. Furthermore, individual actions often serve as tangible expressions of one's values, reinforcing broader cultural shifts toward sustainability.

8.4.2 Sustainable Consumption and Diet

Agriculture remains a leading driver of habitat destruction, whether in tropical rainforests cleared for cattle ranching, or grasslands plowed under for monoculture soy and corn. Reducing meat consumption—particularly beef—can significantly lower one's ecological footprint, as livestock production often demands extensive land and water. A shift toward plant-based diets or moderate meat intake eases pressure on forests and reduces carbon emissions. For those who do eat animal products, choosing pasture-raised or certified sustainable options helps ensure better welfare standards and a smaller environmental footprint.

Seafood choices likewise matter. Overfishing has driven many fish populations to the brink. Opting for sustainably caught or farmed seafood—identified by eco-labels such as the Marine Stewardship Council (MSC)—supports fisheries that adhere to catch quotas, protect bycatch species, and preserve marine habitats. Consumer demand for sustainable fish can encourage retailers and restaurants to stock certified products, creating a supply chain domino effect.

Beyond food, individuals can reduce waste by buying fewer disposable items, choosing products with minimal packaging, and recycling whenever possible. Plastics, in particular, threaten marine life through pollution and ingestion. Replacing single-use plastics with reusable containers or compostable materials cuts down on ocean-bound debris. These shifts may seem minor, but they accumulate when entire communities adopt them.

8.4.3 Energy Conservation and Low-Carbon Lifestyles

Climate change intensifies many drivers of extinction, from coral bleaching to droughts that destroy wetlands. By reducing one's carbon footprint, individuals can mitigate these indirect pressures on wildlife. Installing energy-efficient appliances, weatherproofing homes, and switching to renewable energy sources (where feasible) all contribute. Transportation choices also matter. Walking, cycling, or using public transit lowers emissions compared to private cars, especially for short trips. When longer distances demand air travel, some opt to purchase carbon offsets, though the effectiveness of offsetting programs can vary.

Sustainable living is not limited to wealthy countries. In many developing regions, improved cooking stoves or solar lamps reduce reliance on charcoal or kerosene, curbing deforestation and indoor pollution. Such steps also free up time and resources for families, illustrating how environmental gains can align with improved livelihoods. On a policy level, individuals can urge local governments to support renewable infrastructure, efficient public transportation, and green urban planning.

8.4.4 Supporting Conservation Efforts

Another straightforward yet potent avenue of personal action is the direct support of conservation organizations and projects. Donations, volunteering, or simply spreading the word about NGOs that protect endangered species, restore habitats, or fund research can bolster their impact. Many wildlife reserves and community-based initiatives rely on volunteers to monitor species, clean up habitats, or offer educational programs. Tourists can choose eco-lodges or wildlife-friendly destinations that reinvest part of their revenue into local conservation, helping to sustain biodiversity while delivering economic benefits.

Additionally, individuals can engage politically. Voting for representatives who prioritize environmental policies, signing petitions, attending public hearings on land-use decisions, or contacting elected officials about conservation issues can sway policy outcomes. Writing letters to newspaper editors or participating in social media campaigns fosters public dialogue. Collectively, these actions remind politicians and corporate leaders that biodiversity matters to their constituents and customers.

8.4.5 Cultivating Personal Connections with Nature

Beneath all these pragmatic steps lies a more intangible but profound aspect: rekindling our personal relationships with the natural world. Modern lifestyles often insulate people from the immediacy of ecosystems.

Setting aside time for outdoor activities—birdwatching, hiking, gardening, or simply observing local wildlife—can rebuild empathy and respect for other species. Parents and educators can nurture this connection in children by encouraging them to explore creeks, woodlands, or city parks. The sense of wonder that arises from watching a pollinator feed on a flower, or a frog call at dusk, underpins a lifelong commitment to conserving such marvels.

This reconnection, in turn, shapes our daily choices. People who feel intimately linked to nature are more likely to reduce plastics, conserve water, or question products linked to deforestation. Such individuals also become ambassadors, influencing peers and neighbors to adopt more sustainable habits. In many ways, the most enduring conservation ethic springs not from guilt or fear but from a genuine love of wildlife and wild places—a recognition that preserving Earth's diversity enriches human life and ensures our planet remains vibrant for generations to come.

Conclusion: Aligning Efforts for a Sustainable Future

The magnitude of the extinction crisis can seem overwhelming. Each day's headlines might bring news of another species lost, another forest razed, or another coral reef bleached. Yet against these harrowing signals stands a coalition of conservationists, scientists, local communities, educators, policy-makers, and ordinary citizens who are mobilizing to protect what remains. This chapter has outlined strategies spanning from high-level policy frameworks, such as international treaties and national legislation, to community-led reserves and ex situ breeding. We have seen how restoring habitats—from forests and wetlands to coral reefs—can revive not just individual species but entire ecological processes. We have examined the pivotal role of education in shaping values and fostering stewardship, and we have emphasized that each person's choices—about food, energy, waste, and political participation—can steer society toward or away from sustainability.

If there is a unifying thread to these diverse efforts, it is the understanding that conservation is not a luxury but a necessity. Biodiversity is the infrastructure of life itself, and every extinction chips away at the resilience of our shared planet. While the scientific community refines methods of captive breeding, habitat restoration, and ecosystem management, the social dimension of conservation remains just as critical. Effectively engaging local people, respecting indigenous rights and knowledge, cultivating empathy for non-human life, and nurturing global solidarity form the social scaffolding that can sustain environmental policies and ecological restoration in the long term.

Equally important is the recognition that no single solution can solve the crisis alone. Conservation success stories often emerge from integrating multiple approaches: strong legal protections combined with community empowerment, habitat restoration guided by scientific research, demand reduction campaigns buttressed by local law enforcement, and so forth. Synergy among these components amplifies each one's impact. If a protected area is well-funded but fails to involve local communities, poaching may persist. If ex situ conservation rescues a species in captivity but the original habitat remains degraded, reintroduction might fail. Only a network of interlinked strategies, adapted to local contexts and underpinned by broad-based public support, can effectively counter the pressures driving species to extinction.

The path to a sustainable future is thus a mosaic of top-down and bottom-up initiatives, of grand legislative frameworks and humble everyday acts of care. In continuing to refine and expand these actions, humanity might yet turn the tide. We have barely begun to tap our collective ingenuity and willingness to cooperate. From small farmers practicing agroforestry that maintains pollinator corridors, to global accords that limit greenhouse gas emissions and protect migratory routes across continents, countless examples show that transformation is possible. And while the scars of past extinctions cannot be undone, they can serve as cautionary tales that spur us toward more enlightened stewardship of the living world.

In moving forward, we must recall that the question "What can we do?" is not purely rhetorical. It invites each of us to reflect on our talents, resources, responsibilities, and passions. A university student might focus on research or campus activism, while a farmer might pioneer wildlife-friendly methods. A corporate executive could shift supply chains to sustainable sources, and a policy-maker might champion stronger environmental regulations. The sum of these efforts, multiplied across societies worldwide, underpins our best chance of preserving Earth's wondrous diversity. By acting with intention, we affirm a vital truth: the story of extinction, though grave, is not yet finished. Through concerted efforts, creativity, and compassion, we can write a future where life in all its forms thrives alongside humanity, enriched rather than imperiled by our presence on this beautiful and fragile planet.

CONCLUSION

The story of life on Earth is inseparable from the story of change. Over billions of years, species have arisen, diversified, and sometimes vanished in the turbulence of natural processes—plate tectonics, climatic shifts, asteroid impacts. Yet never before have extinctions proceeded as swiftly or on such a vast scale due to a single species: Homo sapiens. Over the course of this book, we have surveyed the litany of modern extinctions across mammals, birds, reptiles, amphibians, fish, and insects. We have considered not just the biological voids left by these vanished creatures, but also the human factors—hunting, habitat destruction, pollution, disease, climate change—that collectively drive the "sixth mass extinction." We have explored how these losses reverberate through ecosystems, undermining biodiversity and threatening essential ecological functions. And we have examined pathways of hope: the tools, strategies, policies, and daily actions that can slow or even reverse the tide of extinction.

In this final section, we turn our gaze to the future, to the question of what lies ahead for biodiversity and how humanity can fulfill its responsibility to preserve the living tapestry of our planet for generations to come. Although the challenges are formidable, they do not overshadow the capacity for innovation, cooperation, and moral growth that characterizes our species at its best. These pages outline key themes that can guide our efforts in the decades ahead.

1. The Unfinished Story of Biodiversity

We often speak of the "diversity of life" as if it were a static treasure, something that might be simply catalogued and preserved in its current state. In truth, biodiversity is dynamic and ever-evolving. The disappearance of a species ends one branch of the evolutionary tree, but new branches continue to develop in ongoing processes of adaptation and speciation. At the same time, unprecedented human pressures now dwarf many natural evolutionary pathways; habitats once relatively stable for millennia have been fractured or degraded within a single generation. Even if new species occasionally emerge or adapt to urban or altered environments, the rate of anthropogenic change is so accelerated that it outstrips the capacity of most wild populations to cope.

This dynamic tension suggests that our efforts must aim not just to "freeze" the diversity that remains, but to safeguard the ecological and evolutionary processes upon which new diversity relies. Protected areas and wilderness reserves function best when they allow for large-scale natural processes—fire cycles, predator-prey interactions, pollination networks—to continue. Restoration efforts that reintroduce keystone species (such as top predators or large herbivores) can reactivate evolutionary pressures that have been absent for decades or centuries, reviving the potential for complex ecosystems to evolve in place. By understanding biodiversity as a living, fluid tapestry, conservation shifts from a merely reactive stance to one that embraces and facilitates ecological and evolutionary dynamism.

2. Hope Through Collective Responsibility

Central to the future of biodiversity is the notion of **collective responsibility**. If extinctions have accelerated because of humanity's expansion, then humanity must also be the driving force behind protection. This recognition is not new; for decades, conservationists, indigenous communities, and environmental scholars have underscored the moral and practical imperatives of living within ecological limits. What distinguishes our era is the urgency and the scale. The world's human population has surpassed eight billion. Economies are tightly interconnected, meaning that choices about land use or consumption in one region can influence forests, fisheries, or wildlife thousands of miles away. Climate change, a product of worldwide greenhouse gas emissions, intensifies habitat loss and spurs extreme weather events that further endanger species.

Against this backdrop, the notion of personal and societal responsibility becomes multifaceted. National governments, local communities, corporations, non-governmental organizations, and individuals each have roles to play, though with varying degrees of influence. While large-scale policy shifts—limiting deforestation, regulating fisheries, setting aside marine and terrestrial protected areas—are crucial, so too are grassroots initiatives that empower local stewards. A village that sustainably manages its local forest or reef, harnessing ecotourism revenue and respecting traditional ecological knowledge, can serve as a powerful model. A city that invests in green infrastructure—urban parks, tree corridors, pollinator-friendly plantings—can support wildlife diversity even in a heavily built environment.

The global discourse also draws upon ethical frameworks. Religious traditions, philosophical treatises, and indigenous cosmologies often emphasize kinship with non-human life. Such values can galvanize conservation action, especially when aligned with evidence-based strategies. Where once the conversation might have been limited to scientific or economic arguments for saving species, there is a growing recognition that moral and spiritual insights can deepen public resolve. After all, biodiversity conservation is not just about preserving "resources" for human use; it is about recognizing the intrinsic worth of life's variety and our obligation to sustain the conditions that allow it to flourish.

---

3. The Imperative of Climate Action

As we anticipate the future of biodiversity, no factor looms as large or as urgent as **climate change**. Rising temperatures, shifting precipitation patterns, ocean acidification, and more frequent extreme weather events are reshaping habitats across the planet. Coral reefs, for instance, face the dual threats of warming waters and acidification, while terrestrial species must contend with new climatic zones that may emerge too rapidly for them to migrate or adapt. The result can be mass mortalities—like coral bleaching or forest die-offs—or gradual declines in reproductive success, fertility, and distribution ranges.

Tackling climate change is thus a core component of preventing further extinctions. The Paris Agreement, aimed at limiting global temperature increases to well below 2°C above pre-industrial levels, represents one concerted international effort. Yet current emissions trajectories remain perilously high, and many scientific models suggest that a 2°C warming may still be too severe for some ecosystems. Mitigation efforts—transitioning away from fossil fuels, promoting renewable energy, reforesting degraded lands to sequester carbon—must accelerate. Adaptation measures will also be essential. These include creating wildlife corridors to enable species to move, protecting climate refugia such as high-altitude forests or deep-sea habitats, and actively managing habitats to sustain key ecological processes under changing conditions.

A challenge lies in ensuring that climate solutions themselves do not harm biodiversity. Large-scale hydropower dams can fragment rivers, wind farms can disrupt bird or bat populations if poorly sited, and some fast-growth monoculture plantations (used for carbon offsets) can deplete local biodiversity. A synergy approach is needed: carbon capture methods that also enhance wildlife corridors, renewable energy sited away from migratory hotspots, and ecosystem-based climate adaptation that fortifies natural defenses (like mangroves for coastal protection) while maintaining habitat richness. These integrated strategies embody the concept of "nature-based solutions"—the idea that healthy ecosystems serve both climate and biodiversity goals.

4. Innovations in Science and Technology

The modern era brings not just challenges but also an array of scientific and technological tools unimaginable a few generations ago. Genomics and CRISPR-based gene editing allow conservation biologists to probe the genetic health of threatened populations and, in some controversial cases, consider "de-extinction" or assisted evolution to bolster species' resilience. Drones, satellite imagery, and AI-driven analytics can monitor remote habitats, track migrating animals, and detect illegal activities such as poaching or logging. Advances in aquaculture, hydroponics, and synthetic biology might reduce the need for destructive practices in wild areas.

Yet these breakthroughs also raise questions about the ethics and practicality of "high-tech conservation." Should we rely on gene editing or cloning to rescue species that could otherwise be preserved by safeguarding habitat? Might a focus on glamorous technological fixes overshadow the more fundamental need to reduce habitat destruction, pollution, and unsustainable consumption? The key, many argue, is to employ technology as a complement rather than a substitute for robust ecosystem-based strategies. Data-driven approaches can guide law enforcement against wildlife crime or refine habitat restoration, but they cannot replace the political will to set aside land, limit emissions, or respect indigenous rights. Balancing innovation with humility remains a crucial challenge for the conservation community.

5. Education and Engagement: Sustaining Momentum

As discussed in Chapter 8, education, advocacy, and community involvement anchor long-term conservation success. Yet the scale of transformation needed goes far beyond a single generation's schooling. Children who grow up appreciating the wonder of biodiversity may carry these values into adulthood, influencing careers and consumer choices. At the same time, older generations with entrenched habits must also adapt. Hence, a continuous process of learning and engagement must involve schools, media outlets, civil society organizations, and cross-cultural exchanges.

Increasingly, conservation is recognized as a transdisciplinary field. Ecologists and wildlife biologists collaborate with sociologists, economists, educators, and spiritual leaders to craft holistic interventions. Citizen science thrives in the digital age: smartphone apps that identify birds or insects connect amateurs and experts, crowdsourcing data on species distribution. Local ecological knowledge merges with satellite monitoring to identify poaching hotspots. Religious institutions host tree-planting drives or highlight scriptural teachings on caring for creation. Collectively, these endeavors help normalize conservation as a shared social enterprise, rather than a niche concern for specialists.

---

6. Economic Realignment: From Exploitation to Stewardship

A significant driver of extinction is the longstanding economic paradigm that treats nature as an infinite resource or a backdrop for human production. Wildlife habitats are cleared for short-term profit—be it timber, livestock, or mineral extraction—while the broader societal and ecological costs go unaccounted for. The future of biodiversity hinges on transitioning to an economic model that recognizes natural systems as foundational capital, essential for climate regulation, soil fertility, and the intangible benefits we derive from contact with thriving ecosystems.

This shift might involve new valuation metrics: ecosystem service payments, carbon credits tied to forest preservation, or pollination markets that reward farmers who maintain habitat for bees and other pollinators. Such mechanisms attempt to align financial incentives with ecological stewardship. Yet caution is needed. Some critics argue that commodifying nature can reduce it to a transactional basis, overshadowing intrinsic values. Others note that without strong regulatory frameworks and social equity safeguards, market-based conservation can benefit wealthy corporations or landowners at the expense of marginalized communities.

In an ideal scenario, economic realignment also involves supporting small-scale producers who adopt agroecological methods, rotating crops, practicing integrated pest management, and preserving habitat corridors for wildlife. Local cooperatives can pool resources to market certified sustainable products, ensuring premium prices that reflect ecological care. Consumers in distant cities can choose fair trade, organic, or biodiversity-friendly goods, effectively voting with their wallets. These linkages can create virtuous cycles: as demand for sustainability rises, more producers shift practices, generating further biodiversity gains.

---

7. Intergenerational Responsibility

Perhaps the most profound dimension of extinction is its permanence. When a species disappears, the unique evolutionary history and genetic heritage it embodies are lost forever. This irreversible outcome underscores our moral duty to act as responsible stewards for the benefit of future generations. The concept of **intergenerational responsibility** holds that we do not inherit the Earth merely for our own satisfaction, but rather borrow it from those who will come after us. The biodiversity tapestry, which has taken aeons to assemble, is entrusted to our care for a fleeting moment.

This perspective resonates with many ethical and cultural traditions. It calls upon us to weigh not only present-day economic or political pressures but also the well-being of children yet unborn. In practical terms, intergenerational responsibility translates into long-range planning: designating protected areas that remain inviolate for centuries, supporting scientific research that might bear fruit for future breakthroughs, preserving seed banks and genetic repositories, and championing environmental policies that extend beyond short electoral cycles. It also means grappling with climate change at the scale needed to prevent catastrophic temperature rises, so that people fifty or a hundred years from now do not inherit a fundamentally diminished planet.

---

8. Charting a Path Forward

The dilemmas explored in this book—whether about the extinction of high-profile mammals like tigers and rhinos or the quiet disappearance of insects critical for pollination—collectively highlight the fragility of life's interconnections. They also emphasize that solutions are possible where vision, courage, and cooperation prevail. We can point to success stories: whales rebounding after commercial whaling bans, wolves reintroduced to landscapes where they once roamed free, wetlands restored to bring back waterfowl and amphibians. While these successes remain scattered relative to the global scope of the extinction crisis, they prove that positive change is not an illusion; it is a real, if uneven, process.

In forging a hopeful future, a few guiding principles may help unify our efforts:

1. **Holistic Conservation**: Treat species not in isolation but within their ecological contexts. Protect habitats at the landscape or seascape scale, ensuring ecological processes—fire regimes, migrations, seed dispersal—are maintained.
2. **Equity and Inclusion**: Recognize and respect the rights of indigenous and local communities, whose stewardship has historically maintained

high levels of biodiversity. Conservation that empowers rather than displaces local people fosters enduring results.

3. **Adaptive Management**: Environments are not static, particularly under climate change. Policies and management plans must be flexible, adjusting to new data, emerging threats, and evolving scientific insights. Adaptive management also implies continuous monitoring to track outcomes and refine strategies.

4. **Innovation Balanced with Restraint**: Embrace genetic research, remote sensing, AI-driven analytics, and other breakthroughs as tools that enhance (rather than replace) fundamental conservation work—habitat protection, legislation, and community engagement.

5. **Moral Imperative**: Ground conservation in ethical convictions: that non-human life has intrinsic worth, that we owe future generations a living planet, and that the tapestry of life is valuable beyond mere economic metrics.

6. **Global Solidarity**: Biodiversity loss does not respect national borders. Many species migrate across countries or occupy transboundary ecosystems. International coordination, fair resource-sharing, technology transfer, and financial support for biodiversity-rich but economically challenged regions are all crucial.

7. **Cultural Transformation**: Expand education and outreach so that conservation becomes second nature for societies everywhere. Cultivate a planetary ethic that sees humanity as part of nature, not above it.

---

9. A Shared Destiny

In the end, the crisis of extinction beckons us to reconsider our place in the web of life. Are we mere exploiters, extracting resources until they run dry? Or can we be caretakers, forging a sustainable coexistence that honors the diversity and wonder of existence? The potential for either outcome lies within us, shaped by the choices we make as societies and as individuals.

The pressing reality is that every extinction chips away at Earth's capacity to sustain vibrant ecosystems—ecosystems upon which we, too, depend. The comforting notion that human ingenuity will "save us" regardless of what happens to other species misses the deeper point: we are part of these ecosystems. When they unravel, our cultural, emotional, and material well-being unravels with them.

Yet the future is not scripted. Just as human societies can devastate habitats, we can also restore them. Just as we can drive species to extinction, we can also pull them back from the brink. This book's journey through the lost wonders of mammals, birds, reptiles, amphibians, fish, and insects should serve not as a eulogy, but as a clarion call. Each species' story, from the thundering herds of

Quaggas to the solitary call of a vanished bumblebee, reminds us of nature's abundance, its fragility, and our capacity both to destroy and to conserve.

Perhaps our most profound responsibility is to ensure that future generations have the chance to learn from living creatures rather than just museum specimens. That they can marvel at the songs of birds at dawn, the iridescent flash of a butterfly's wings, the roar of lions, the deep hush of a tropical forest breathing at dusk. By embracing multi-faceted strategies—conservation policy, habitat restoration, technology, education, sustainable lifestyles—and uniting them under a shared moral commitment, we can maintain and rebuild the richness of life on Earth.

In stepping forward, let us remember that hope is not naive optimism. It is a willingness to see the challenges clearly, yet still choose to act. It is a conscious decision to align our lives, our communities, and our economies with the flourishing of life in all its forms. If we succeed, the next chapters in the story of Earth's biodiversity will celebrate resilience rather than mourn loss. Our descendants will look back on this pivotal moment not as the dawn of an irreversible extinction crisis, but as the turning point when humanity chose to uphold its responsibility as steward, guardian, and co-inhabitant of this extraordinary, ever-evolving planet.

www.ingramcontent.com/pod-product-compliance
Lightning Source LLC
LaVergne TN
LVHW101613210926
848480LV00008B/489
* 9 7 9 8 2 3 0 8 1 7 4 7 5 *